Nanobotany

Sumera Javad • Ayesha Butt
Editors

Nanobotany

Springer

Editors
Sumera Javad
Department of Botany
Lahore College for Women University
Lahore, Pakistan

Ayesha Butt
Department of TESOL
Lahore College for Women University
Lahore, Pakistan

ISBN 978-3-030-08374-8 ISBN 978-3-319-77119-9 (eBook)
https://doi.org/10.1007/978-3-319-77119-9

Softcover re-print of the Hardcover 1st edition 2018

Printed on acid-free paper

This Springer imprint is published by the registered company Springer International Publishing AG part of Springer Nature.
The registered company address is: Gewerbestrasse 11, 6330 Cham, Switzerland

Dedicated to our parents
whose efforts and sacrifices have enabled us
to do something in life

Preface

This book is a link between botany and nanotechnology. This book is intended to fill the gap in interrelated fields of applied research for botany students.

Since last 100 years, there is an increased interest of research groups and enhanced tendency of investments in the fields of nanotechnology. Nanotechnology has really revolutionized the era and has greatly speeded up the progress of technology. Plant sciences or botany researchers have found their interest in the field of nanotechnology and once again, this renewable green treasure of nature has proved its vitality for the bulk production of nanoparticles. So, this book is an effort to link both the fields. No doubt, nanotechnology is a very broad term and field, but this study is an effort to understand the nature of nanoparticles and then practically find their applications in botany.

In this era of fast growing technology, there is need for bulk production of nanoparticles. So, there is a concept of this production with the help of plant metabolites. There are details of mechanisms which can be opted to develop nanoparticles with controlled size and shape with plant metabolites. It is a special effort by different scientists from various fields of life.

It is expected that this book will be equally resourceful for the botany students as well as for the nanotechnologists. This book will have an academic impact for graduate level and postgraduate level students as well. Chapter 1 and 2 will give an idea of plants to nanotechnologist group. Chapter 3 to 5 and 10 may help the botanist group to understand the insight of technology while all other chapters are directing both groups to reach to the roots of the link. This book will definitely be changing the vision of botany students.

Lahore, Pakistan — Sumera Javad

Contents

List of Figures

List of Plates

List of Tables

Chapter 1
Plants and Nanoworld: an Introduction

Nadia Ghaffar and Sumera Javad

1.1 Introduction

This Century ended the difference of basic fields like mathematics, biology, physics etc. and converted them to inter related applied fields like biophysics, biochemistry, biostatistics etc. It paved the way of progress to all the modern technologies available for human beings. These developments in interdisciplinary fields helped scientists in their research and reduced the time of invention magically.

Metal nanoparticles occupy now a basic status in technology as well as in medicines. They are being synthesized by a number of chemical and physical methods. Physical methods include the use of laser and microwaves etc. These methods will be described in detail in Chap. 3 of this book. These methods involve a lot of chemicals, energetic waves, man power and time. There is definitely need of such a method which relies on some renewable source of energy, some natural source with lesser by products as an input for synthesis of nanoparticles. Considering these points we come to ever green treasure of nature i.e. The Plants (Fig. 1.1).

Study of plants is known as Botany. Plants include all flowering (Angiosperms) and nonflowering plants (Gymnosperms, pteridophytes, bryophytes). This is the field of science describing the physiology, ecology, genetics, metabolism, cytology and chemistry of plants on surface of earth. It is one of the oldest knowledge related to science (Fig. 1.2).

Plants grow in all types of environments. They can tolerate extreme environmental conditions like water stress and temperature stress etc. Their physiology and metabolism is changed according to the environment around. Even it has also been reported that different plants produce different types of biochemicals during different seasons i.e., winter and summer. Nature of biochemical produced by plants is also affected by the type of soil in which they are present. It is also dependent upon

N. Ghaffar · S. Javad (✉)
Department of Botany, Lahore College for Women University, Lahore, Pakistan

S. Javad, A. Butt (eds.), *Nanobotany*, https://doi.org/10.1007/978-3-319-77119-9_1

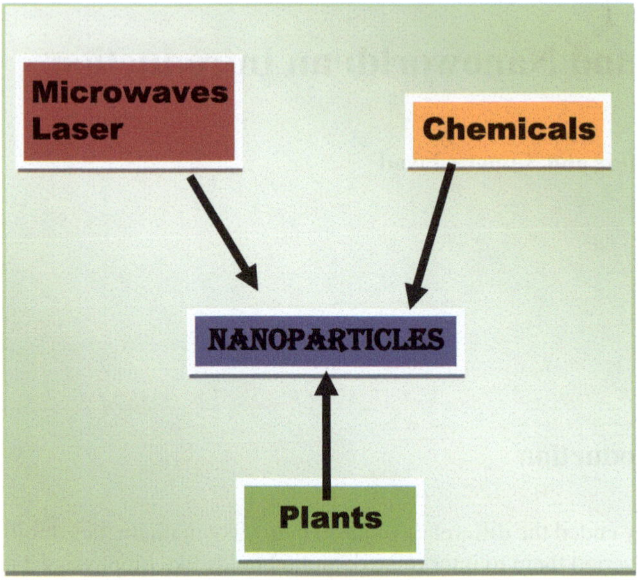

Fig. 1.1 Common sources to synthesize nanoparticles

the genetic makeup of the plant. It means what types of genes are present in that species. So when we talk about the plant based chemicals, we have a tremendous variety and a few of which are described in Chap. 2 of this book.

Nanotechnology is the study of technology of nanoparticles. Nanoparticles are the particles with at least one dimension less than 100 nm. Literature is filled with the information on practical applications of nanoparticles. Researches are now being carried out on development of a sustainable method for the bulk production of nanoparticles. Nanoparticles need some antioxidants or reducing agents whether natural or chemicals or waves, for their synthesis from metal salts and plant metabolites take the responsibility here.

So here arises another practical aspect of applied botany which can be named as "NanoBotany" as it refers to the nano-sized particles synthesized by plants or related to plants. NanoBotany is future of interdisciplinary applied sciences.

1.2 Plant Metabolites

A substance essential to the metabolism of a particular organism or to a particular metabolic process is called metabolite. Human beings have relied upon plants for their necessities like food, shelter and cloths. History shows that they have used different parts of plants for this purpose like leaves, wood, bark, buds, fruits etc. They were also used for arrow poisoning, hallucinations for rituals, poisons for murders and for hunger depressions etc. (Harborne 1984). Nearly 80% of the world's

Fig. 1.2 A Gymnosperm

population relies on traditional medicines for primary health care, most of which involve the use of plant extracts. In India and Pakistan, majority of the prescriptions are plant based in the traditional systems of Unani, Ayurveda, Homeopathy and Siddha (Sandhya et al. 2006).

A plant cell produces two types of metabolites, first are primary metabolites which are produced to target growth and development. Examples of primary metabolites are carbohydrates, proteins and fats while secondary metabolites are produced other than primary metabolites as a byproduct of the plants' main metabolic pathway. They are not basically required for plant growth and development directly. These secondary metabolites don't have any importance in plant classification as they are produced randomly. But only one thing to remember is that they are not junk. They have a purpose in nature. Several secondary metabolites produced in plants play many important functions. For example, terpenoids inhibit competing plants, protect plants from insects and attract pollinators. Alkaloids help plants to resist herbivores and insects. Plant secondary metabolites also help in cellular signaling. Concentrations of theses secondary metabolites may be different in different plants and in different parts of plants but they are playing their role (Bernhoft 2010).

It has also been purposed that plant secondary metabolites are typically defense chemicals. They may be terpenoids, alkaloids, resins or essential oils, they may be specie specific, area specific, family specific but are produced usually in small quantities and it's a difficult job to isolate them (plantscience4u).

These secondary metabolites are now being worked out to be used in nanobiotechnology. They are very good antioxidants naturally and have a great potential to reduce the natural oxidants produced during plant metabolism. They too have a very good potential of forming metal nanoparticles as explained in Chapter no 2 of this book. These may be considered as the very good source of raw material for bulk synthesis of nanoparticles. For practical applications of nanoparticles, they should be of controlled shape and size. It has also been suggested in a number of studies that by controlling the pH, temperature and time of reaction of metal salt with the plant extract.

1.3 Classes and Number of Secondary Metabolites

There is a variety of plant secondary metabolites associated with variety of plants. These secondary metabolites may be specie specific or family specific. Their quantity may vary with environmental stresses. Detail of number of these secondary metabolites is given in Fig. 1.3.

Based on their biosynthetic origins, plant secondary metabolites can be divided into three major groups:

1. Terpenoids
2. Nitrogen-containing alkaloids and sulphur containing compounds.
3. Flavonoids and allied phenolic and polyphenolic compounds.

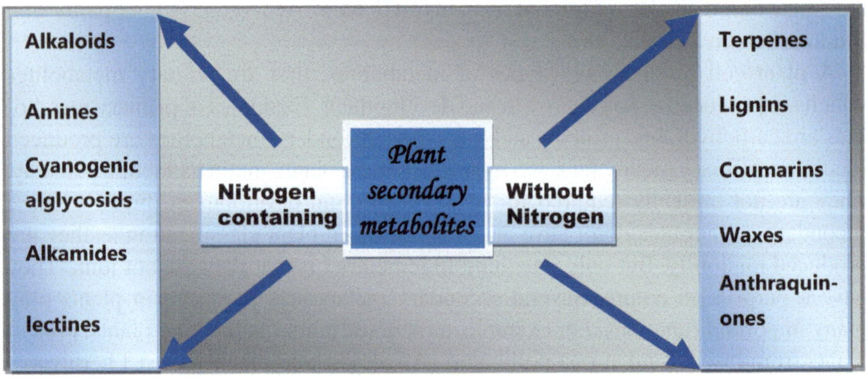

Fig. 1.3 A few types of plant secondary metabolites

1.4 Terpenoids

Terpenoids are the plant secondary metabolites with very diverse nature, like they may be linear in structure or poly cyclic or aromatic. They may be 5-C hemiterpene or having very much complex structure like rubber. But whatever the terpene is, it is made up of the condensation of isoprene units. Isoprene units are formed by the acetate by Mevalonic acid pathway which then join head to head to form different classes of terpens (Mahmoud and Croteau 2002). Terpens are classified by the number of isoprene units they have in their structure.

Some of the common Terpenoids are Steroids, carotenoids, and gibberelic acid, showing its diversity. There are more than 2300 known structures of Terpenoids. They are part of essential oil composition. They are extensively used in the industries, like perfume industry, cosmetics, flavors and spice industry. A number of Terpenoids have medicinal importance having pharmacological activity (Styger et al. 2011).

1.4.1 Mono and Sesqui-terpenoids, and Phenylpropanoids

These are low molecular weight volatile compounds. They are diverse and found among all plant species. Monoterpenes consist of two isoprene units while sesquiterpenes have three units. Phenylpropanoids have less known structures but they form the relative groups consisting of only nine carbons. Their synthetic pathway is also different. These metabolites are lipophilic in nature and are volatile compounds i.e., having low boiling points. They give strong taste and smell. They are extremely used in herbal remedies. Their medicinal uses include antibacterial, antiviral, antifungal and antineoplastic etc. However if these are concentrated then these have toxic effects. The plant family rich in these compounds is *Lamiaceae*, the thyme family. They are also found in other families (Bernhoft 2010).

1.4.2 Diterpenoids

Diterpenoids are consisting of four isoprene units and usually they are not volatile. They don't have any odor. They are lipophilic, so give taste. They have very less known toxocoligical effects. They are found in number of plant families and have very well known medicinal uses (Bernhoft 2010) (Fig. 1.4).

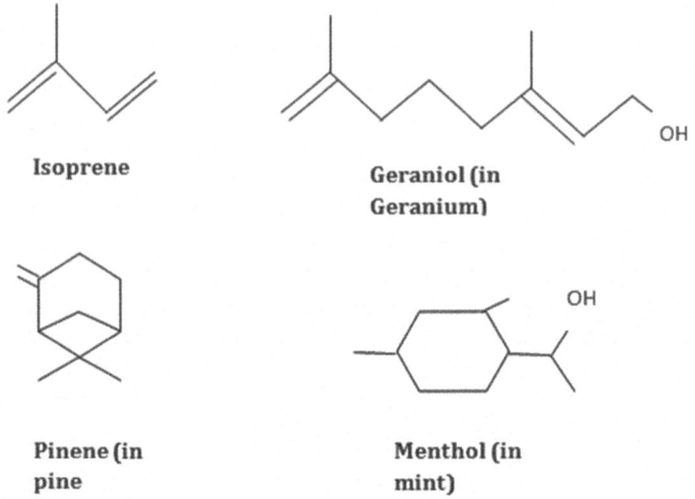

Fig. 1.4 Structure of some terpenoids

1.4.3 Functions

Plants tissues are very much known for their capability to adapt to both biotic and abiotic stresses such as herbivores, pathogens, water and temperature etc. Terpenoids are key to this. However, some terponoids may also affect the surroundings of these plants because they are added to the environment due to their high volatile and reactive nature. Terpenoids also act as infochemicals because they are sensed by the pollinators, neighboring plants and by herbivores, so helping the sessile plants to live. Physiologically function of plant volatiles or terpenoids can be defined in three ways,

1. Plant-plant interaction,
2. The signaling between symbiotic organisms
3. Attracting pollinators i.e., insects.

These roles of terpenoids make them an agricultural tool for pest control to an industrial application for production of flavors and fragrances (Maffei 2010).

1.4.4 Plant Families Rich in Terpenoids

Lamiaceae
Rubiaceae
Cannabaceae
Ginkgoaceae
Lauraceae

1.5 Alkaloids

Alkaloids are basically nitrogen containing heterocyclic compounds, bitter in taste with potent activity. Alkaloids also include some compounds having neutral and weakly acid properties. Besides nitrogen, carbon and hydrogen, alkaloids also contain sulfur, oxygen and rarely some other elements like chlorine, phosphorus and bromine. A clear cut differentiation between alkaloids and nitrogen containing natural compounds is not well known. Some other compounds such as amino acids, amines, peptides, proteins, nucleic acid and nucleotides are not commonly called alkaloids. Their distribution in the plant kingdom is limited but some groups contain various clinical properties (Giweli et al. 2013).

1.5.1 Families Rich in Alkaloids

(i) **Tropane alkaloids** are produced in *Solanaceae* (nightshade family) for example *Hyoscyamus niger* (henbane), *Datura* spp. (thorn apples) and *Atropa belladonna* (deadly nightshade). The compounds usually contain anticholinergic activity (muscarine receptor antagonists) so, used for the treatment of hypersecretion, pain and smooth muscle spasms.

(ii) **Pyrrolizidine alkaloids** are found in *Asteraceae* (daisy family), mostly in *Boraginaceae* (borage family) and *Senecio* spp. (Ragworts). After bioactivation they cause hepatotoxicity in man and animals.

(iii) **Isoquinoline alkaloids:** The main producers of **isoquinoline alkaloids** are *Berberidaceae* (barberry family) and *Papaveraceae* (poppy family). They have biochemical effects and used for medical purpose. They can be used for the inhibition of bacteria, cancer cells and pain, for myocardial contractility, and to stimulate bone marrow leucocytes.

(iv) **methylxanthine alkaloids:** *Theobroma cacao* (cacao) and *Coffea arabica* (coffee) produce **methylxanthine alkaloids**. Binding of Methylxanthines to adenosine receptor causes neurological effects in animals and man, low to moderate intake causes stimulation. High intakes of methylxanthines in rodents elicit reduction in sperm production and testicular atrophy.

(v) **Pseudoalkaloids** are found in families *Apiaceae* (carrot family) and Taxaceae (yew family) for example *Conium maculatum* (hemlock) and *Cicuta virosa* (cowbane); *Taxus baccata* (yew) respectively. Chemical properties of pseudoalkaloids are closely related to alkaloids. *Conium maculatum* and *Cicuta virosa* produce some pseudoalkaloids that greatly effects the central nervous system (Bernhoft 2010).

(vi) **Furocoumarines and naphthodianthrones** Furocoumarines have photosensitizing properties that are mostly produced in *Apiaceae* (carrot family) predominantly in *Heracleum* spp. (cow parsnips). The naphthodianthrones for example in *Polygonaceae* (dock family) e.g. *Fagopyrum esculentum* (buckwheat) and in *Hypericum* spp. (St. John's-worts) of *Clusiaceae* (garcinia family) are similar in action (Fig. 1.5, Table 1.1). Mostly antidepressant effects have been observed in *Hypericum* spp.

1.6 Phenolic Compounds

Plants contain secondary metabolites which have many functional groups. One of them is phenolic group made up of aromatic ring and hydroxyl group. Approximately 8000 different phenolic structures have been studied in plants. Phenols are very diverse in nature and may present in the form of single ring to complexed polyphenols and tannins. Phenols are mostly found attached with carbohydrates and other organic compounds. Classification of these compounds can be done on the number and symmetry of carbon atoms. Common two types of phenolic compounds are flavonoids and non-flavonoids. (Strack 1997) (Fig. 1.6).

1.6.1 Flavonoids

Most common and largest amount of phenolic compounds in the form of polyphenols are flavonoids in plants. These flavonoids are found in epidermis of leaves and fruit's outer covering to protect the plant from diseases, UV absorption, pigmentation and provide stimulus for nitrogen fixing in the nodules of plants. Structurally, flavonoids have 15 carbons and two benzene rings linked with 3 carbon atoms. The main subclasses of flavonoids are the flavones, flavonols, flavan-3-ols, isoflavones, flavanones and anthocyanidins (Koes et al. 1994).

Some other less common flavonoid groups i.e. coumarins, dihydroflavanols, dihydrochalcones, aurones and flavan-3, 4-diols are also found in our diet. Basic structure of flavonoids has hydroxal group attached to fourth, fifth and seventh carbon. Carbohydrates i.e. glycosides are conjugated with flavonoids are naturally occurring. These glycosides and hydroxyl groups enhance the water solubility of flavonoids. Methayl and isopentyl units in flavonoids made them lipid loving in nature. Flavonols are mostly found in vegetables, fruits and beverages but a difference in their sizes has been seen because of environmental and varietal changes (Crozier et al. 1997).

1.6.2 Flavones

Flavones have two aromatic rings which are attached to heterocyclic pyrene ring and have a base of 15 carbon atoms. Different types of flavones have also been seen i.e. flavonols, flavanone, flavones and many others. Flavonoids are classified into

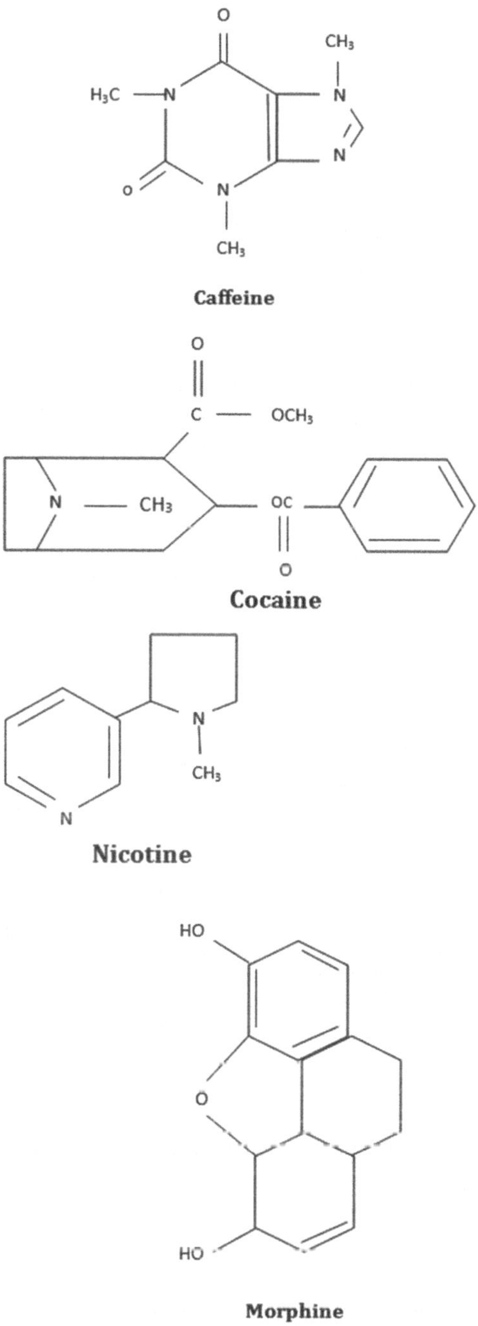

Fig. 1.5 Examples of alkaloids, a diverse group of secondary metabolites that contain nitrogen (Taiz and Zeiger 2010)

Table 1.1 Pharmacological effects of some well known alkaloids

Alkaloid name	Source	Pharmacological activity	References
Nicotine	*Solanaceae* plants family	Stimulant, antiherbivore, insectide, anti inflammatory	Gandhi (2013)
Morphine	*Papaver somniferum* and poppy derivatives	Act on CNS (central nervous system), on myenteric plexus, acute pulmonary edema and reduce the shortness of breath	Rozov-Ung et al. (2014)
Coniine	*Conium macularum, Sarracenia flave*	Neurotoxin, poisonous	Panter et al. (2013)
Atropine	*Atropa belladonna, Darura stramonium, Mandragora officinarum Berberis species, Hydrastis, Canadensis*	Competitive antagonist of muscarinic acetylcholine receptors, anti cholinergic, anti myopia effects	McBrien et al. (2013)

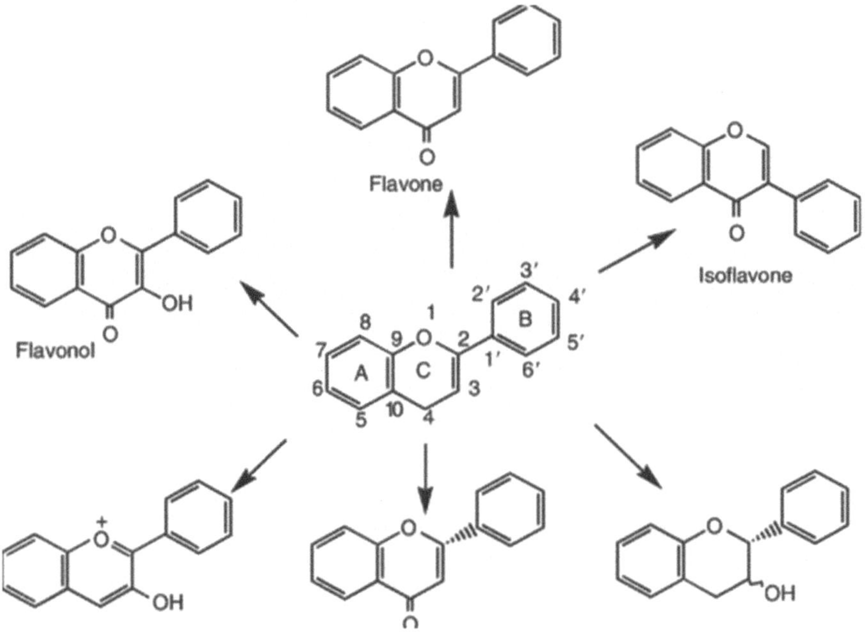

Fig. 1.6 Generic structure of major flavonoid (Anurag et al. 2015)

different classes upon the position of carbon atom and oxidation level whereas compound differences within class is due to pattern of two aromatic rings.

Flavonols are closely associated with flavones in structure. Some Flavones like apigenin and luteolin do not have oxygenation at C_3 but they switch positions with A and C ring.

Flavones have diverse nature for substitution i.e. glycosylation, methylation, alkylation and hydroxylation. Most flavones occur as 7-O-glycosides. Flavones are not much common and found only in significant plants such as parsley, celery and in specific herbs. Some polymethoxylated flavones in citrus plants have also been reported. Some flavones are also associated with diseases such as millet flavones cause goitre in West Africa (Gaitan et al. 1989).

1.6.2.1 Flavan-3-ols

Flavonoids have structurally complex type which include Flavan-3-ols and have simple monomers i.e. (+)-catechin and its isomers to more complexed polymeric condensed form known as tannins. These complexed proanthocyanidins, flavan-3-ols and flavanones are non-planar and have C_3 in heterocyclic ring while other types of flavonoids i.e. tannins, flavonols and many other have planar molecules. Flavan-3-ols have two central chiral carbon atoms C2 and C3 which forms 4 isomers for each step of B-ring hydroxylation. Two of these isomers are (+)-catechin and (−)-epicatechin are common in nature while other two (−)-catechin and (+)-epicatechin are rare (Clifford 1986).

1.6.2.2 Isoflavones

Isoflavones are abundant in leguminaceae especially in soya bean which have 5 main classes of isoflavone. Structure of isoflavone is categorized by the presence of B ring linked to C_3 instead of C_2 position.

1.6.2.3 Plant Families Rich in Phenolic Compounds

Cannabaceae
Ericaceae
Lamiaceae
Pedaliaceae
Rosaceae

1.7 Tannins

Tannin is natural polyphenols and its name derived from a French word Tanin means tanning substance. The tannins are diverse group of oligomers and polymers and they can synthesize complex with minerals, proteins, cellulose and starch. The synthesizes pathway of phenolics such as isoflavons lignins and aromatic amino acids similar to the formation of phenloics and this pathwayis known as shikimic acid pathway, also called as the phenylpropanoid pathway. Tannins are subdivided into two groups based on their solubility in water named as PA (Proanthocyanidins), also known as condensed tannins but with the exception of high molecular weight molecules.

HT (hydrolysable tannins) that comprises elligatannins, gallotannins and complex tannins contain active principles of plant based medicinal so tannins having plants are used as a diuretic against stomach, astringents against diarrhea, and anti-inflammatory and duodenal tumors according to literature (Brito 2007).

1.7.1 Plant Families Rich in Tannins

Leguminosae
Anacardiaceae
Combretaceae
Rhizophoraceae
Myrtaceae
Polinaceae

1.8 Glycosides

Glycosides consist of two moieties one is sugar moiety and other is non-sugar moiety so they may be alcohol, or sulfur compounds or phenol. Many plants store chemicals in form of inactive glycosides and when plants need those stored glycosides, enzymes hydrolyze them into active compounds, e.g. inactive glycosides are subgrouped as prodrugs, they are hydrolyzed to produce aglycone which is active part of glycosides. Glyycosides are catagorised into cardiac glycosides, anthraquinone glycosides, glucosinolates, saponins and cyanogenic glycosides (Polt 1995) (Fig. 1.7).

1.8.1 Categories of Glycosides

The **cardiac glycosides** hydrolyze to release aglycone portion which have a steroidal structure. Aglycone part has inhibitory effect on Na^+/K^+ ATPase pumps in the cell membranes. These ATPase pumps maintain the concentration of ions in cardiac cells so is very pronounced in the heart functioning resulting the contractility and reduction in heart beating. Plants that have cardiac glycosides are of *Convallariaceae* (convall family) with *Convallaria majalis* (lily of the valley) *Scrophulariaceae* (figwort family) particularly *Digitalis purpura* (foxglove).

The **cyanogenic glycosides** have amino acids as aglycone moiety. Many compounds of such glycosides have function in iodine consumption and result in hypothyroidism. The other lethal effect is the release of hydrogen cyanide at high dosages. Plants of families that are rich source of cyanogenic glycosides are *Rosaceae* (rose family) especially in *Prunus* spp.

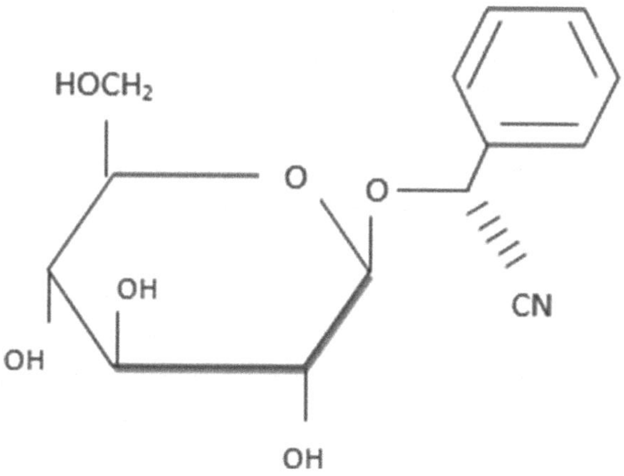

Fig. 1.7 Structure of glycoside (James 2011)

The **glucosinolates** contain aglycone portion consist of pungent sulphur-containing amino acid. Such glycosides indicate a set of effects on cytochrome P450 as tend to reduce hepatic bioactivation of environmental procarcinogens in isoforms found in many cells. These compounds can cause hypothyroidism and goiter and can also induce skin irritation. The plant family is mostly associated with the production of such glycosides is *Brassicaceae* (brassica family).

Saponins mean "soap forming compounds" and mostly they are found as glycosides. The non-sugar moiety aglycones comprises of either tetracyclic steroids or pentacyclic triterpenoids. They have different structure but similar in functions as aglycone portion is hydrophyllic in nature and glycine part is hydrophobic which have soapy features. This is the reason that they are used in detergents. Saponins have antineoplastic effects and show immune modulating. Saponins cause haemolysis of Rbcs *in vitro*, while such effects are not visible *in vivo*. Some saponins cause photosensitisation and jaundice. Saponins are common in a number of plant families. *Liliaceae* (lily family) is rich in having saponins with particular example of *Narthesium ossifragum* (bog asphodel) which is also known as sheep toxic plant.

Anthraquinone glycosides These glycosides are not very diverse among the kingdom of plants. We find their examples from family *Polygonaceae* (dock family) in the form of *Rumex spp.* (curly dock) and *Rheum* spp. (rhubarbs). They are very well known to produce water and electrolyte secretion along with peristalsis in colon (Bernhoft 2010).

1.8.2 Families Rich in Glycosides

Rosaceae
Liliaceae
Polygonacea
Brassicaceae
Convallariaceae

1.9 Conclusion

Plants produce bioactive compounds via secondary metabolic pathway which are known as secondary metabolites. These metabolites are produced in very small amount and are not directly involved in growth of plant but perform many different functions and may elicit many different effects in animals and humans. Plants with such bioactive compounds contain both medicinal and adverse effects which depends upon the amount consumed and type of plant species. Food plants with such bioactive compounds with smaller amount of effects are usually regarded as beneficial and used for pharmacological effects. These secondary metabolites have strong antioxidant potential which can be used as a natural method to produce nanoparticles particles by reduction of metal ions from their metal salts.

References

Anurag K, Anumalik Y, Nitika G, Swadesh K, Nikhil G, Santosh K, Vinay Y, Anuj P, Himanshu G (2015) Metabolites in plants and its classification. World J Pharm Pharmcal Sci 4(1):287–385

Bernhoft A (2010) A brief review on bioactive compounds in plants. Department of Feed and Food Safety, National Veterinary Institute, P.O.Box 750 Sentrum, NO-0106 Oslo, Norway

Brito AM (2007) Synthesis and characterization of glycosides. Springer, Cham

Clifford MN (1986) Phenol protein interactions and their possible significance for astringency. In: Birch GG, Lindley MG (eds) Interaction of food components. Elsevier Applied Science, London, pp 143–164

Crozier A, Lean MEJ, McDonald MS (1997) Quantitative analysis of the flavonoid content of commercial tomatoes, onions, lettuce and celery. J Agr Food Chem 45(1):590–595

Gaitan E, Lindsay RH, Reichert RD (1989) Antithyriodic and gioterogenic effects of millet: role of C-glycosyl flavones. J Clin Endocrin Metabol 68:707–714

Gandhi PT (2013) Novel nicotine derivatives US Patent, 20130123106

Giweli AA, Dzamic AM, Sokovic M, Ristic M, Janackovic P, Marin P (2013) The chemical composition, antimicrobial and antioxidant activities of the essential oil of *Salvia fruticosa* growing wild in Libya. Arch Biol Sci 65(1):321–329

Harborne JB (1984) Phytochemical methods: a guide to modern techniques of plant analysis, 2nd edn. Chapman and Hall, New York

http://plantfacts.osu.edu/resources/hcs300/biochem3.htm

http://www.plantscience4u.com/2013/02/primary-and-secondary-metabolites.html#.U4sKh85L - zA

http://www.tankonyvtar.hu/en/tartalom/tamop425/0010_1A_Book_angol_01_novenyelettan/ch03s05.html

https://knoji.com/medicinal-plants-what-makes-them-different-from-other-plants

James RC (2011) Medicinal plants: what makes them different from other plants. Natural guide to medicinal herbs and plants, Frantisek Stary

Koes RE, Quattrocchio F, Mol JNM (1994) The flavonoid biosynthetic pathway in plants: function and evolution. BioEssays 16:123–132

Maffei M (2010) Sites of synthesis, biochemistry and functional role of plant volatiles. South Afr J Bot 76(4):612–631

Mahmoud SS, Croteau RB (2002) Strategies for transgenic manipulation of monoterpene biosynthesis in plants. Tren Plant Sci 7:366–373

McBrien NA, Stell WK, Carr B (2013) How does atropine exert its anti myopia effects? Ophthal Physiol Opt 33(3):373–378
Panter KE, Welch KD, Gardner DR, Green BT (2013) Poisonous plants: effects on embryo and fetal development. Birth Defects Res C Embryo Today 99(4):223–234
Polt RL (1995) Method for making amino acid glycosides and glycopeptides, U.S. Patent No. 5,470,949. U.S. Patent and trademark office, Washington, DC
RozovUng I, Mreyoud A, Moore J, Wilding GE, Khawam E, Lackner JM, Sitrin MD (2014) Detection of drug effects on gastric emptying and contractility using a wireless motility capsule. BMC Gastroenterol 14(1):2
Sandhya BS, Thomas WI, Shenbagarathai R (2006) Complementary and alternative medicines 3:101–114
Strack D (1997) Phenolic metabolism. Plant biochemistry. Academic, London, pp 387–416
Styger G, Prior B, Bauer FF (2011) Wine flavor and aroma. J Indus Microorgan Biotech 38(9):1145–1159
Taiz L, Zeiger E (2010) Secondary metabolites in plant defense. Plant Physiol 22:395

Chapter 2
Advanced Concept of Green Synthesis of Metallic Nanoparticles by Reducing Phytochemicals

Zubaida Yousaf and Nadia Saleh

2.1 Introduction

There is an air of chaos and insecurity in developing countries about the contemporary techniques of agricultural productivity and an advanced way of dealing every day life problem. Nanotechnology in present era is an exciting discipline of science for production of metallic nanoparticles by reduction of metal ions. It resulted with the existing products leading to the formation and alteration of profitable products. This field of research is opening new horizons of investigation and getting more appreciation in agriculture, industry and food sector (Maynard et al. 2006; Rejeski and Lekas 2008; Dawson 2008; Elumalai et al. 2011; Husen and Siddiqi 2014).

There is an expectation of about more than 15% intrusion of nanotechnology products by the next years (Husen and siddiqi 2014). This will be a game changer and will build an innovative way of dealing every trouble by merger of science and technology arenas such as electrical engineering, mechanical engineering, supra molecular chemistry, colloidal sciences, material sciences, applied physics, pharmaceutical and agricultural sciences. Nanotechnology is serving in the fields of food industry, non linear optical devices, optics, gene delivery, health care, tissue engineering, biomedical and many more (Mohanpuria et al. 2008). The world turning era for use of nanoparticles in the field of drug delivery was 1950s and the 1960s. Professor peter Paul research group was first to investigate polyacrylic beads for oral administration. They got success in developing first nanoparticles for drug delivery and vaccines. This discovery lead toward unveiling tremendous uses of nanoparticles not only for drug delivery but for development of delivery systems, transportation of drugs across the blood brain barriers (Kreuter 2007). New techniques like TEM, AFM, and DLS have uplifted the nanotechnologies to the stages where it is considered as the future to all technologies (Singh et al. 2010). Now a

Z. Yousaf (✉) · N. Saleh
Department of Botany, Lahore College for Women University, Lahore, Pakistan

S. Javad, A. Butt (eds.), *Nanobotany*, https://doi.org/10.1007/978-3-319-77119-9_2

days nanotechnology has applications in designing novel drugs, delivery systems and devices in the field of biomedical sciences.

A number of chemical and physical approaches are used by the scientists to synthesize nanoparticles but these are proven to be highly toxic and hazardous for environment. Alternative approaches like green synthesis or biogenic/biomimetic synthesis of nanoparticles are chic now a days. The method of phytosynthesis of nanoparticles by phyto reducing agents is eco friendly, cost efficient, relatively reproducible, one step, less time consuming, efficient, sustainable and has productive edge, at room temperature and pressure (Mathur 2014). Green synthesis is the scale up of nanoparticles from plant and plant parts and isadvantageous over the other biological resources like fungi, bacteria and actinomycetes. Green synthesis of nanoparticles from plant extract or phytochemicals is less expensive, safer, biocompatible, readily reproducible and easily achievable in comparison to whole plant (Wiley et al. 2005; Roy and Barik 2010; Govindaraju et al. 2010). The contemporary approach to synthesize phyto nanoparticles from phytochemicals is much simpler approach.

2.2 Role of Phytochemicals in Nanoparticles Synthesis

Phytochemicals are bioactive compounds which reduce the risk of diseases in our body. Non nutritive phytochemicals are secondary metabolites such as terpenoids, tannins, flavonoids, steroids, alkaloids and many more which perform antioxidant activity by protecting our body from free radicals released from exposure to radiation, tobacco smoke or other pollutants. These free radicals cause oxidative damage in our cell wall, lipids, DNA, RNA, proteins and enzymes. Medicinal plants are rich of these antioxidants, which have ability to neutralize free radicals and to combat against oxidative breakdown of cells (Peteros and Uy 2010; Agbafor and Nwachukwu 2011; Sies et al. 1992). The properties of the nanoparticles sythenesised by green technologydepend upon the nature and source of the plant extract or phytochemicals. Different sources, natureand concentration of phytochemicals in plant extract determine the structure and properties of nanoparticles. This is the key step to synthesize nanoparticles of desirable quality for the targeted problem by manipulating plant metabolites to a certain extent and concentration. Phytochemicals act as reducing as well as capping and stabilizing agents at once. Green production of nanoparticles is through very simple technique, whichproceeds as plant extract containing phytochemicals is simply mixed with metallic salt at room temperature at specific concentration, incubation time, electrochemical potential and pH. This process takesfew minute for completion (Dwivedi and Gopal 2010; Li et al. 2011). At the start of green chemistry, plant selection was based on hit and trail methods. Then it was further improved by the use of medicinal and aromatic plants extract due to the presence of active and readily available phytochemicals in them (Mathur 2014). The most common, water soluble/polar solvent phytochemicals are alkaloids, terpenoids and phenolics compounds. Non protein part of enzymes also reported to have reduction potential for metallic salts (Mittal et al. 2013). The properties of metallic nanoparticles are shape and size dependent (Jiang et al. 2009). The most commonly used metallic salts are gold, silver, zinc, copper and iron (Mittal et al. 2013; Husen and Siddiqi 2014). There is a need to quantify the phytochemicals of plants to determine their effectiveness and potential to form nanoparticles(Vasanthi et al. 2014). These phytochemicals can be a efficient source to reduce metal salts in synthesis of nanoparticles.

Recently identified phytochemicals such as flavonoids, terpenoids, flavones, ketones, aldehydes, amides and carboxylic acid have the greater potential to reduce nanoparticles(Jha et al. 2009). Immediate reduction was recorded by water soluble phytochemicals like quinones, organic acids and flavones. *Cyprus sp.* (Mesophytes), *Hydrilla sp.* (Hydrophytes) and *Bryophyllumsp.* (Xerophytes) were found to be involved in reduction of sliver nanoparticles. In xerophytes emodine and anthraquinones were undergone radical tautomerization which resulted in synthesis of sliver nanoparticles. In mesophytes activation of quinones leads the reduction of silver nanoparticles. Transformation of catechol into protocatechaldehyde and then to protocatecheuic acid was responsible for the reduction of nanoparticles of 2–5 nm (Jha et al. 2009). Phytochemicals are mostly present in seeds, fruits, leaves and stem. Green synthesis is a strong bridge between the plant sciences and nanotechnology serving for the human welfare either in form of nanomedicines or biomedical field (Dekker and Verketrk 2003; Nishino et al. 2007; Johnson 2007; Holst and Williamson 2008).

2.3 Nanoparticles Synthesis from Plant Parts and Whole Plant

Plants are rich sources of heavy metals (photosynthate) that can be used in the biosynthesis of nanoparticles. There are a number of plants which have been reported for the synthesis of gold and nanoparticles. Alfalfa (*Medicago sativa*) plant was used for the synthesis of gold nanoparticles by providing it targeted metal salt rich environment(Iravani 2011; Gardca-Torresdey et al. 1999, 2000, 2002a, b, 2003). Plants contain the surplus amount of bio molecules which can be used to reduce metal ions for the synthesis of nanoparticles in single step. Synthesis of nanoparticles from whole plant extract has been studied. The method requires more effort for significant results and synthesis of nanoparticles due to ample amount of bio molecules in whole plant extract. Shape and size of the nanoparticles cannot be manipulated due to varietyofphytochemicals inthe plant extract. The minor changes in thephysical and chemicalconditionsduringnanoparticlesbio synthesis canmodifytheend product. With the advancement of knowledge use of plant extracts from plant parts and tissues (leaves, barks, seeds, fruits, roots and others) have received much more appreciation.

However the use of whole plant or *in vivo* synthesis of nanoparticles is a much slower process and expensive than plant extract (Park et al. 2011; Mubarak et al. 2011; Daisy and Saipriya 2012; Mukunthan and Balaji 2012; Silva et al. 2015) Interestingly, different plant extracts of whole plant and plant organ have different concentrations of phytochemicals and pH. Furthermore, pH value of plant extract has greater influence in reduction of metal ions which is mediated by concentration of their natural phytochemicals. Plant extracts having small reducing ability are supplied with exogenous supplementation (such as proteins, peptides and amino acids) to boost the process (Vijayakumar et al. 2013; Makarov et al. 2014). Plants use their primary and secondary metabolites for their physiological responses to stress conditions and surviving strategies. There can be variations in the phytochemicals due to the change in seasons, climate, reproductive behavior, pathogen and herbivores. The metabolites vary in concentration and composition according to biotic and abiotic conditions of plant. Use of whole plant extract or fraction extract changes the nanoparticles morphology and chemistry. These conditions affect the reproducibility and scaling up of metallic nanoparticles synthesis. So there is a hard need to develop much simpler and optimized technique for the biosynthesis of nanoparticles, which may be reliable and consume lesser energy (Mittal et al. 2013).

2.4 Reproducibility of Nanoparticles

High reproducibility of nanoparticles is a challenging matter these days. There are several factors that affect the reproducibility of metallic nanoparticles while synthesis with plant extracts. This is due to the reason that plantshave to face a number of drastic changes like environmental stressincluding water, temperature and light exposures, herbivoresor parasitic attack, species competition, plant ageandsoil pH. These all factorscauselack orexcessof essential nutrients. The incidental pressures are also there like pollution, pesticide spray and anthopogenic actions. The factors which can affect scale up of the nanoparticles, are low abundance of plant material, lack of appropriate equipment for necessary amount, variation in molecular composition and heterogeneous plant extracts. These stimuli regulate concentration and composition of plant metabolites. Genetic factor and physiology of plant also greatly influence the metabolite regulation of plant (Ali-Shtayeh et al. 1997; Verpoorte et al. 1999; Wang et al. 2000; Lisar et al. 2012). If the composition and concentration of a plantextract are different then they produce different sizes of nanoparticles. For the high reproducibility rate of nanoparticles time of collection of plant material is an essential requirement. There are many scientists who did not pay attention to the period of collection and quantification of metabolites in the plant extract. However, these factors are necessary to culminate the synthesis of metallic nanoparticles. Elucidation of molecular mechanism can lead to devise the way to produce an optimized composition for the synthesis of nanoparticles by combining two or more plant extracts.

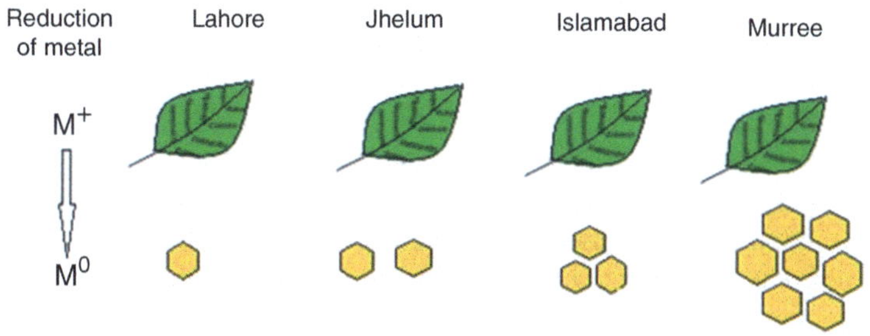

Fig. 2.1 Pictorial presentation of nanoparticles synthesized from leaves of same plant species collected from different regions having different conformation and morphology

The other approach is the regulated growth of plant in controlled conditions and environment or producing plant by tissue culture or GMOs technique. The technique will allow the rapid and facile production of nanoparticles by accumulation of plant metabolites in plants organs. The plant organ containing metabolites will be directly available for the metallic nanoparticles synthesis (Verpoorte et al. 1999) Same condition of batch is also needed. Absence or presence of a new plant metabolite can be expected. If the plant extracts used in large amount for nanoparticles synthesis there can be a dramatic difference in the phytochemicals of plant extract of similar plant species. The difference can lead to a large variation in the reducing ability and structure of nanoparticles (Fig. 2.1). There is a foremost need of the scientists to determine the composition of the plant metabolites which are involved in the synthesis of nanoparticles by reducing metal ions. Adequate characterization is a prerequisite to maximize the possibility of plausible results. The current need of green nanoparticles biosynthesis demands predictable, scalable, reproducible and safe metallic nanoparticles product (Rao and Ravishankar 2002).

2.5 Nanoparticles Synthesis from Phytochemicals/Plant Metabolites

Various plant metabolites involved in bioreduction of metallic nanoparticles are terpenoids, alkaloids, phenolics, polyphenolics, sugars, proteins and acids. These phytochemicals reduce metabolites to form nanoparticles differently and produces diverse structures of nanoparticles. Their compositions in plant extracts are the key component which are involved in the synthesis of specific nanoparticles (Makarov et al. 2014). By varying the duration of synthesis, the yield, physical and chemical structure of nanoparticles also have modified the functional character. This creates a problem for the scientist because plant extracts do not always have same composition. It varies with the change in ecological and climatic factors. There comes a need to optimize the solution that is involved in synthesis of metallic nanoparticles. For the optimization plant extract must be quantified to maintain the concentration of metabolites that are involved in bioreduction. The step will increase the reproducibility factor of the nanoparticles (Makarov et al. 2014).

It is concluded from the large amount of experimental work for green synthesis of nanoparticles that plant extract or green synthesis in the most reliable, feasible, cost effective and eco friendly method. It is advanced technique to resolve problems of mankind either in biomedical field or in engineering fields. The synthesis must be desireable, reproducible, targeted and stable without causing any toxicity. Different plant extract contains different composition of plant metabolites in it which affect the entire morphology, physiology and chemical structure of the nanoparticles and can also affect the results. It is reported that different phytochemicals behave differently in synthesis of matallic nanoparticles. To cope the difference in desired results and for the optimized and standardized results, proper identification and isolation of phytochemicals is needed, which is still under consideration and needs further investigation (Park et al. 2011; Ahmed et al. 2016).

2.5.1 Nanoparticles Synthesis from Terpenoids

Terpenoids are the diverse organic polymers,synthesized from 5-carbon isoprene units. It was demonstrated that substance possess antioxidant activities, is involved in synthesis of nanoparticles involved in antioxidant activity (Hu 2011). Terpenoids are helpful in bio reduction of silver nanoparticles in geranium leaves. *Cinnamomum zeylanisum* (cinnamon) contains a terpenoid named as eugenol found to be the major component in bio reduction of silver and gold nanoparticles (Shiv Shankar et al. 2003). The proton dissociated from the eugenol leaves the OH- group behind which is involved in further oxidation accompanied with reduction of metallic nanoparticles (Fig. 2.2). Red color in compound shows presence of functional group at the position involved in reduction of nanoparticles. Khan et al. (2016) reported that terpenoids are the rich source of essential oils act as one of the best capping and reducing agent for the synthesis of uniform shaped and uniform sized nanoparticles.

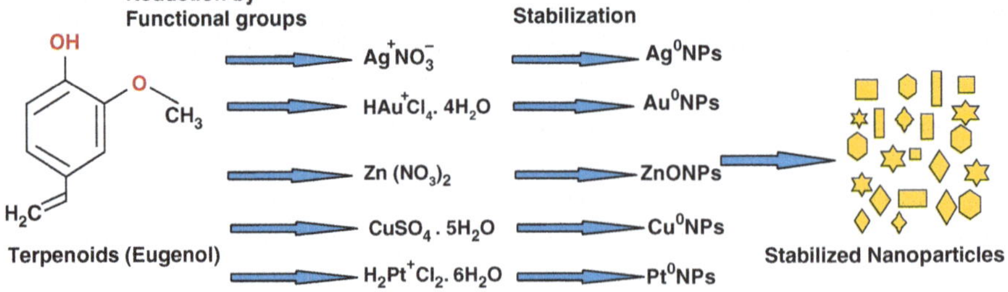

Fig. 2.2 Synthesis of various shapes of nanoparticles from different metallic ions by using terpenoids (Eugenol)

Terpenoids are oxygenated forms of all the terpenes comprised of 90% of essential oils. They can be in form of monoterpenoids, diterpenoids, hemiterpenoids and sesquiterpenoids. Although monoterpenoids and sesquiterpenoids are major building blocks for the synthesis of silver nanoparticles (Sangwan et al. 2001; Khan et al. 2016). Plant species like *Anacardium occidentalis*, *Myristica fragrans*, *Ricinus communis*, *Cocos nucifera* and vegetables containing terpenoids are direct source of formation of stable nanoparticles (Zwenger 2008; Da Silva et al. 2008; Bakkali et al. 2008; Zamiri et al. 2011; Sheny et al. 2012; Kumari and Philip 2013). Different concentration and forms of terpenoids are responsible for synthesis of diverse shape and size of nanoparticles which are further involved in several bio medical activities such as antibacterial, antimicrobial and cytotoxicity. Its potential to reduce and stabilize silver nanoparticles makes it a liable reducing agent (Vilas et al. 2014). Conclusively, terpenoids are the most important phytochemicals for nanoparticles synthesis and direct use of these compounds can fill the gap to get most stable and targeted yield. It will allow high compositional homogenity of nanoparticles and it will be an effective way to increase the chance of reproducibility of nanoparticles (Gengan et al. 2013).

2.5.2 Nanoparticles Synthesis from Flavonoids

Flavonoids are the polyphenolic group of plant metabolites comprising anthocyanins, isoflavonoids, flavones, chalcones, flavonols and flavanones. It is postulated from the studies that the tuatomeric transformations of the flavonoids from enol to keto group were involved in bio reduction of metallic nanoparticles. For example, there has been reported bio synthesis of nanoparticles from flavonoids of *Ocimum basilicum* named rosmarinic acid and luteolin. Hydrogen released during transformation of enol to keto group involved in bio reduction of metallic ions (Singh et al. 2010). Further conversions of ketones to carboxylic acids are responsible for gold particles reduction. Quercetin is a flavonoid has a good chelating ability due to the presence of hydroxyls and carbonyl groups at C5 and C3 sites. It can chelate several metallic ion such as Fe^{3+}, Fe^{2+}, Zn^{2+}, Cu^{2+}, Cr^{3+}, Al^{3+}, Co^{2+} and Pb^{2+}. The phenomena explains the adsorption ability of the flavonoids on the nascent nanoparticles. The mechanism depicts their role in nucleation (initiation) of nanoparticles and process of aggregation inclusively bioreduction from plant metabolites (Fig. 2.3). Different classes of nanoparticles were involved in synthesis of different nanoparticles structure. It was reported that a flavonoid from *Lawsonia inermis* (henna) named apiin (apigenin glycoside) help in synthesis of anisotropic gold nanoparticles and quasi spherical silver nanoparticles (21–30 nm in size) (Ahmad et al. 2010). FTIR analysis displayed the apiin attachment to the nanoparticles by carbonyl group. Alkaloids are considered to be the most efficient reducing agents which are effective against many diseases and are used in formation of nanomedicines. Flavonoids were reported to be involved in synthesis of irregular shaped gold nanoparticles from buds of dried clove (*Syzygium aromaticum*) (Kasthuri et al. 2009). The falvonoids were found to

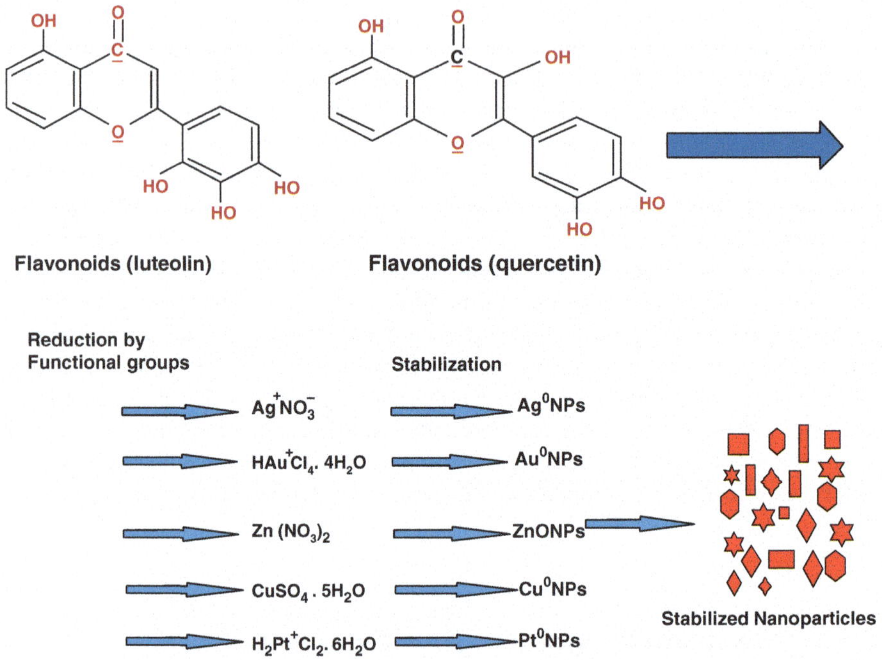

Fig. 2.3 Synthesis of various shapes of nanoparticles from different metallic ions by using flavonoids (Luteolin and quercetin)

be involved in pharmacological activities like antiallergic, anti inflammatory, antiviral, cytotoxic and anti bacterial activities (Murray 1998; Cushnie and Lamb 2005; Kasthuri et al. 2009; Raghunandan et al. 2010). Stephen and Seethalakshmi (2013) reported the use of silver nanoparticles directly from flavonoids "hesperidin" instead of whole plant extract for optimization and easy availability. Flavonoids are considered to be the most active and potential source of various nanoparticles effective against anti carcinogenic and antimicrobial activities.

2.5.3 Nanoparticles Synthesis from Sugars

Sugars can also reduce metallic nanoparticles but their ability to reduce nanoparticles varies with the structure of the sugar. Monosaccharides contain single sugar such as glucose molecule (linear) comprises an aldehyde group involved in reduction process by conversion into keto group. Fructose is a monosaccharide sugar which contains keto group and acts as an antioxidant. The functional group changes the functional ability of metallic nanoparticles. However, disaccharides and polysacharrides ability to reduce nanoparticles varies with the monosaccharides sugars present in the linear chain which provides the access of the nanoparticles to

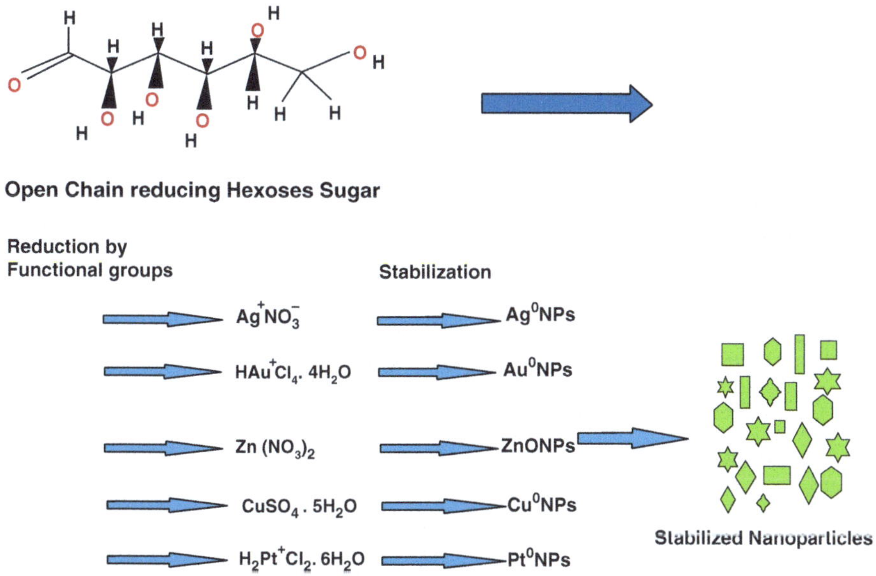

Fig. 2.4 Synthesis of various shapes of nanoparticles from different metallic ions by using open chain reducing hexose sugar

aldehyde group (Cook and Samman 1996). For example, in disaccharides sugars such as lactose and maltose contains at least one sugar, which has linear structure and exposed functional group involved in bioreduction (Fig. 2.4). Furthermore, in case of sucrose there is no functional group available in linear form that can reduce metallic nanoparticles. Due to the acidic hydrolysis of fructose and glucose in linear chain of sucrose metallic salts transformed into tetrachloroauric acid and tetrachloroplatinic acid. Aldehyde group is converted to carboxylic group through OH nucleophilic addition, leads to the metallic ion reduction for the synthesis of nanoparticles. The similar mechanism was observed in gold nanoparticles reduction from magnolia vine extract (Murray 1998). Glucose can synthesize various morphology of nanoparticles due to its stronger reducing character whereas fructose produce only monodispersed symmetry which determines their antioxidant ability is limited to the tautomeric shifs of the functional groups (Cook and Samman 1996).

2.5.4 *Nanoparticles Synthesis from Proteins*

Proteins are the polmeric units of amino acids found to be frequently linked to nascent nanoparticles (Shiv Shankar et al. 2003). Different amino acids have different binding and reducing ability. For example, amino acids like cysteine, lysine, methionine and arginine have ability to reduce silver nanoparticles (Zayed et al. 2012). Aspartate is capable to reduce tetrachloroauric acid, whereas lysine and valine lack

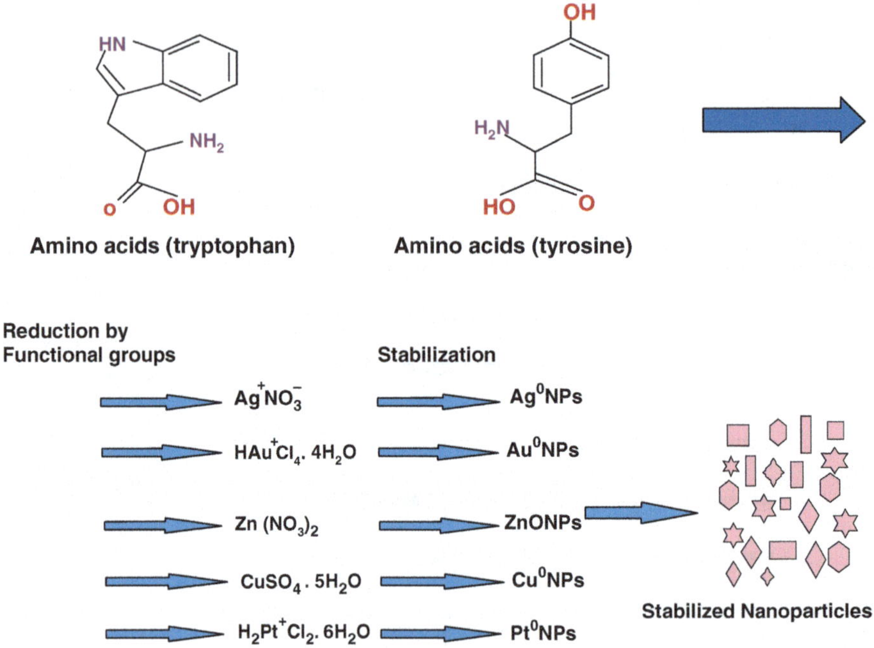

Fig. 2.5 Synthesis of various shapes of nanoparticles from different metallic ions by using amino-acids (Tryptophan and tyrosine)

this ability (Njagi et al. 2010). The reducing and binding ability of 20 natural alpha amino acids was determined. It was observed from the studies that tryptophan has the strongest reducing ability of gold ions whereas histidine is considered to be strongest binding agent for gold nanoparticles. The metallic ions bind with the amino acids either through carboxylic or amino group such as in imidazole ring of the histidine attachment is through nitrogen atom and in glutamic or aspartic acid binding is with carboxylic group. There are other examples of sides group involved in binding thioether (methionine), thiol (cysteine), carboxylic group (glutamine and asparagine) and hydroxyl (threonine, serine and tyrosine) (Mandal et al. 2002). It was reported that carbonyl group of glutamine and asparagine, hydoxyl group from tyrosine, thiol group from cysteine are responsible the bio reduction of nanoparticles (Fig. 2.5). The linkage of the peptide bonds in amino acids changes the binding and reducing ability of metallic ions. Free chains expose the sites for reduction and binding of nanoparticles. The amino acid sequence also affects the ability and accessibility of functional group and changes the nature of the nanoparticles. The amino acids which have weak binding with metal ions are ineffective in reduction of tetrachloroauric acid anions due their inability to retains near the reduction site, such as proline, phenylalanine and leucine can weakly bind metal ions (Lukman et al. 2011). It is demonstrated from recent studies that the sequence of amino acids in proteins varies the morphology, size and amount of nascent nanoparticles (Tan et al. 2010).

Therefore, it is necessary for the present era to quantify and isolate the plant metabolites which specify the morphology and functional ability of nanoparticles to get the targeted results. This would be optimized and revolutionized approach in every field of science if scientists would be able to manipulate the nanoparticles synthesis for desired results. However, variation in results may be seen due to certain factors. There are some factors which are uncontrollable like climatic factors. This is the strategy to produce the solution that have high reducing ability, higher yield, desired product which has the high reproducibility rate and scale up ability.

2.6 Mechanism of Metallic Nanoparticles Synthesis

There are three phases involved in the synthesis of nanoparticles from plant extract. First is activation phase in which reduction of metallic ions and nucleation of these reduced ions takes place. Second is the growth phase, in this stepthere is an aggregation of small adjacent nanoparticles into large size nanoparticles. Heterogeneous nucleation and the growth involved in direct formation of nanoparticles leads to further metallic ion reduction referred as Ostwald ripening. The process is accompanied by the thermodynamic stability of metallic nanoparticles. The last stage is the termination stage in the synthesis of nanoparticles (Glusker et al. 1999; Tan et al. 2010). In this phase nanoparticles attain the most energetically stable conformation (Fig. 2.6). The stability of the nanoparticles is greatly influenced by the composition of the plant metabolites or plant extract composition. Stability of the nanoparticles reduced by increasing their surface area. In order to achieve the most stable conformation the morphology of the nanoparticles changes in order to derogate the Gibbs free energy. Duration of growth phase modifies the shapes of the nanoparticles. The nanoparticles amass to form nanoprisms, nanotubes, nanohexahedron and a number of irregularly shaped nanoparticles (Mandal et al. 2002; Si and Mandal 2007). The functional groups usually involved in green synthesis are -C-N, -C-O-C, CH_3-C-CH_3, C-H, -OH, CH_3, CH_2, C-O, -C=O, C-C, C=C (Fig. 2.6) (Khan et al. 2016) (Plates 2.1 and 2.2).

2.7 Potential Plants (Phytochemicals) for Synthesis of Nanoparticles

Medicinal and aromatic plants are rich sources of active plant metabolites that can be used in the synthesis of the nanoparticles to combat diseases. In this chapter, our focus is on the phytochemicals which have greater potential to reduce and bind metallic ions and which are most favorable for the stabilized nanoparticles synthesis. There is a list of medicinal plants which contain active secondary metabolites (flavonoids, terpenoids, alkaloids and tannins). They are effective against microbial

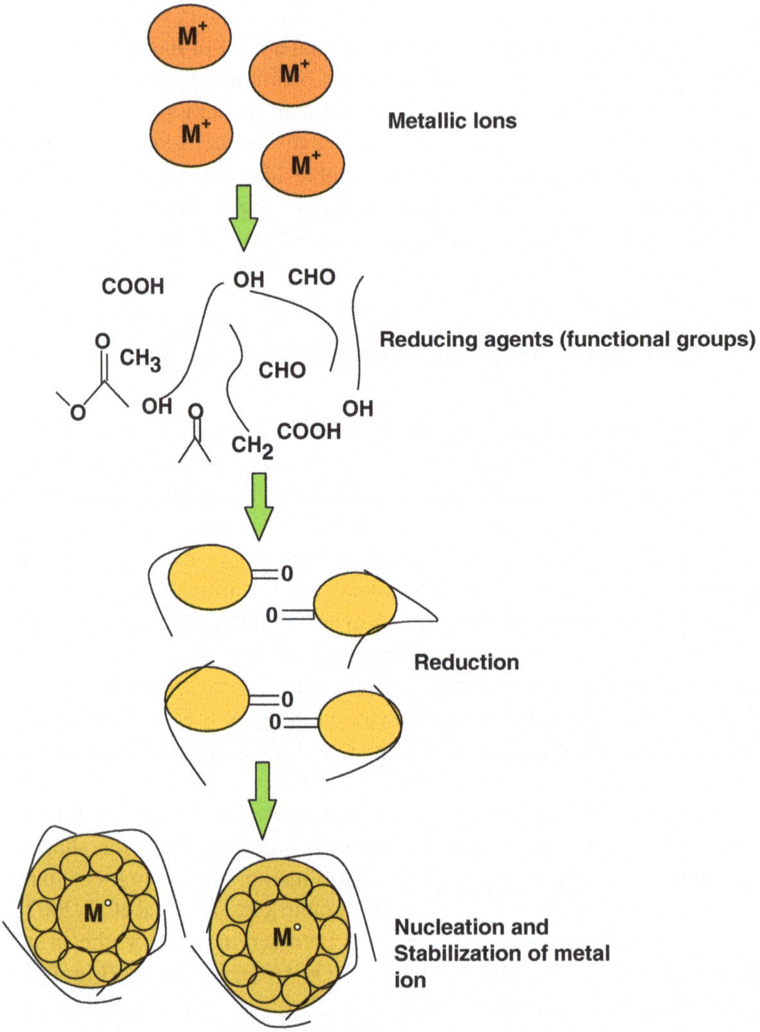

Fig. 2.6 Schematic representation of nanoparticles synthesis from phytochemicals

activity, act as healer and protect against many infectious conditions (Smitha et al. 2009).

There is a list of edible plants comprising active plant metabolites which have higher anti carcinogenic ability. Apples contain flavonoids (anthocyanins, epicatechin and quercetine) and terpenoids. Blackberries, blueberries, raspberries and strawberries are rich source of Elligatannins, Flavonoids (anthocyanins, catechins, kaempferol, quercetin), pterostilbene and resveratrol. Broccoli, arugula, cabbage, Brussels sprouts, cauliflower, collard greens, horseradish, kale, mustard greens, radish, rutabaga, turnip and watercress comprise phytochemicals such as carotenoids

Plate 2.1 Plants used for the synthesis of gold particles

(beta catotene), indole, isothiocyanate (allyl isothiocyanate, benzylisothiocyaate, crambene,phenylethylisothiocyanate,sulforaphaneand3-phenylproyliothiocyanate). In coffee plant caffeine, diterpenes and phenolic acids (chlorogenic acid and quinic acid) are rich in amount. Dark green leafy vegetables are comprised of carotenoids

Plate 2.2 Plants used for synthesis of silver particles

(beta carotene, alpha carotene and lutein) and flavonoids. Dry beans and soy are rich of insitol, flavonoids (isoflavones), lignans, polyphenols, protease inhibitors, saponins, sterols, triterpenoids, phenolic acids, protein kinase inhibitors and sphingolipids. Walnut comprises Elligatannins, flavonoids, phenolic acids, phytosterols. Whole grains are rich in flavonoids, phytic acid, lignans, phenolic acid, saponins and protease inhibitors. All these edible plants and fruits are effective against anti carcinogenic activity due to large amount of phytochemicals. By using these plant metabolites we can produce desired nanoparticles with higher reproducibility rate. Despite of using whole plant extracts isolated of necessary active compounds will save the time, lessen the raw material, higher reproducibility, avoid false results and higher the yield. The strategy is to produce more out put with smaller input. Tissue culture technique can also be helpful by manipulating the growth of the desired phytochemical in a plant tissue.

2.8 Conclusion

Metallic nanoparticles have variety of application like drug delivery, imaging, cosmetics, Biosensors, paints, semiconductor, packaging, automobiles. In short they are in use in each and every aspect of life. However chemical synthesis of these particles either by top-down or bottom-up method is very expensive, produce hazardous compound and is time consuming too. To get the enhanced usage of nanoparticle there was space to develop some cheap, ecofriendly and more stable particles production methods. In the last decade people put an emphasis on the green synthesis of nanoparticles due to their fascinating roles for the welfare of mankind. As the plants are in therapeutic use since human being started to acclimatize this planet. Plant are having broad range of phytochemicals like alkaloids, terpenoids, phenols, flavonoids, tannins and quinines etc. which can mediate the synthesis of nanoparticles. Silver and gold nanoparticles were synthesized by using aqueous and ethanol leaf extracts of various plants. A list of plants is given in Table 2.1 which contains active phytochemicals that are considered to be potential and active source of nanoparticle synthesis for desired and beneficial results.

2.9 Future Prospects

The advancement in the field of nanotechnology application may be the solution of the problems faced by the whole plant extract. There can be implementation of slow nanorange carriers, which released manipulated amount of phytochemicals (Kim et al. 2010). The manipulated release of phytochemical into the solution for the synthesis of nanoparticles can be used to produce desired yield and reproducible results. This will save time, cost and will ensure the targeted end product. These techniques are growing in industries, biomedical, genetics and agricultural fields. In future nanotechnology based products will be a major contributor to enhance the life style of human being by producing nanomedicines (Siddiqui et al. 2015; Katti et al. 2009).

In future,the identification and isolation of desired plant metabolites for nanoparticles fabricationcan lead towards the revolutionary advancement in pharmaceutical industry. Different solvents are used for the extraction of various phytochemical which can mediate the synthesis of nanoparticles. Therefore combinations of desired phytochemicals from different phytochemical extraction can be used for green synthesis and they may possess various therapeutical properties. In this nanotechnology

Table 2.1 Plant containing phytochemicals active against antimicrobial activities

Secondary metabolites	Plant names	References
Terpenoids (eugenol)	*Pimenta dioica* (Allspice), *Syzygium aromaticum* (Clove)	Martinez et al. (1996), Jones and Luchsinger (1986) and Katti et al. (2009)
Essential oil (terpenoid)	*Aegle marmelos* (Bael tree), *Ocimum basilicum* (basil), *Laurusnobilis* (Bay), *Barosma setulina* (Buchu)	Cichewicz and Thorpe (1996), Ofek et al. (1996), and Rana et al. (1997)
Alkaloids (Berberine)	*Berberis vulgaris* (Barberry)	Wan et al. (1998)
Terpenoids (Terebinthone)	*Schinus terebinthifolius* (Brazilian pepper tree)	Rana et al. (1997)
Polyacetylene, tannins, Terpenoids	*Arctium lappa* (Burdock)	Rana et al. (1997)
Essential oils (Catechols, eugenol)	*Piper betel* (Betel pepper)	McDevitt et al. (1996)
Alkaloid (Piperine)	*Piper nigrum* (Black pepper)	Jacob et al. (2012)
Tannins (Polyphenols, anthraquinone)	*Rhamnus purshiana* (Cascara sagrada)	Ghoshal et al. (1996)
Salicylic acids (Polyphenols)	*Anacardium pulsatilla* (Cashew)	Berkada (1978)
Essential oils, (Terpenoids, tannins), others	*Cinnamomum verum* (Ceylon cinnamon)	Berkada (1978)
Anthemic acid (Phenolic acid)	*Matricariachamomilla* (Chamomile)	Bose (1958), Hambuger and Hostettmann (1991), and Himejima and Kubo (1991)
Cocaine (Alkaloid)	*Erythroxylum coca* (Coca)	Cichewicz and Thorpe (1996) and Scheel (2016)
Polyphenol (Tannin)	*Eucalyptus globulus* (Eucalyptus)	Jones and Luchsinger (1986)
Alkaloid (Berberine)	*Mahonia aquifolia* (Oregon grape)	Navarro et al. (1996)
Latex Mix of terpenoids, organic acids, alkaloids	*Papaya Carica* (papaya)	Burdick (1971), Osato et al. (1993), and Omulokoli et al. (1997)
Alkaloids and others (Opium)	*Papaver somniferum* (Poppy)	Burdick (1971), Osato et al. (1993), and Omulokoli et al. (1997)
Terpenoid (Caffeic acids, Tannins), polyphenols	*Artemisia dracunculus* (Tarragon)	Wild (1994) and Satrija et al. (1995)
Terpenoid (Caffeic acid), Phenolic alcohol (thymol), polyphenols (tannins), flavones	*Thymus vulgaris* (Thyme)	Kim et al. (2010)

also contributes towards the conservation of plants; the bulk of plantas material is used for the preparation of traditional medicines which can be replaced by nanoparticles synthesized by small amount of plant material.

References

Agbafor KN, Nwachukwu N (2011) Phytochemical analysis and antioxidant property of leaf extract of *Vitex doniana* and *Mucuna pruriens*. Bio Chem Res Int 2011:1–4

Ahmad N, Sharma S, Alam MK et al (2010) Rapid synthesis of silver nanoparticles using dried medicinal plant of basil. Colloids Surf B Biointerfaces 81(1):81–86

Ahmed S, Ahmad M, Swami BL, Ikram S (2016) A review on plants extract mediated synthesis of silver nanoparticles for antimicrobial applications: a green expertise. J Adv Res 7(1):17–28

Ali-Shtayeh MS, Al-Nuri MA, Yaghmour RMR, Faidi YR (1997) Antimicrobial activity of *Micromeria nervosa* from the Palestinian area. J Ethnopharmacol 58(3):143–147

Bakkali F, Averbeck S, Averbeck D, Idaomar M (2008) Biological effects of essential oils–a review. Food Chem Tox 46(2):446–475

Berkada B (1978) Preliminary report on warfarin for the treatment of herpes simplex. J Irish Coll Phys Surg 22:56

Bose PK (1958) On some biochemical properties of natural coumarins. J Ind Chem Soc 58:367–375

Burdick EM (1971) Carpaine: an alkaloid of *Carica papaya*—its chemistry and pharmacology. Eco Bot 25(4):363–365

Cichewicz RH, Thorpe PA (1996) The antimicrobial properties of chile peppers (*Capsicum* species) and their uses in Mayan medicine. J Ethnopharmacol 52(2):61–70

Cook NC, Samman S (1996) Flavonoids—chemistry, metabolism, cardioprotective effects, and dietary sources. J Nutr Biochem 7(2):66–76

Cushnie TT, Lamb AJ (2005) Antimicrobial activity of flavonoids. Int J Antimicrob Agents 26(5):343–356

Da Silva EC, Da Silva MGA, Meneghetti SMP et al (2008) Synthesis of colloids based on gold nanoparticles dispersed in castor oil. J Nanopar Res 10(1):201–208

Daisy P, Saipriya K (2012) Biochemical analysis of *Cassia fistula* aqueous extract and phytochemically synthesized gold nanoparticles as hypoglycemic treatment for diabetes mellitus. Int J Nanomedicine 7:1189–1202

Dawson NG (2008) Sweating the small stuff: environmental risk and nanotechnology. BioSci 58(8):690–690

Dekker M, Verkerk R (2003) Dealing with variability in food production chains: a tool to enhance the sensitivity of epidemiological studies on phytochemicals. Eur J Nut 42(1):67–72

Dwivedi AD, Gopal K (2010) Biosynthesis of silver and gold nanoparticles using *Chenopodium album* leaf extract. Coll Surf A Physicochem Eng Aspects 369(1):27–33

Elumalai EK, Prasad TNVK, Nagajyothi PC, David E (2011) A bird's eye view of biogenic silver nanoparticles and their applications. Pelagia Res Lib 2:88–97

Gardea-Torresdey JL, Tiemann KJ, Gamez G et al (1999) Gold nanoparticles obtained by bio-precipitation from gold (III) solutions. J Nanopart Res 1(3):397–404

Gardea-Torresdey JL, Tiemann KJ, Gamez G, Dokken K et al (2000) Reduction and accumulation of gold (III) by *Medicago sativa* alfalfa biomass: X-ray absorption spectroscopy, pH, and temperature dependence. Env Sci Tech 34(20):4392–4396

Gardea-Torresdey JL, Parsons JG, Gomez E et al (2002a) Formation and growth of Au nanoparticles inside live alfalfa plants. Nano Lett 2(4):397–401

Gardea-Torresdey JL, Tiemann KJ, Parsons JG et al (2002b) Characterization of trace level Au (III) binding to alfalfa biomass (*Medicago sativa*) by GFAAS. Adv Env Res 6(3):313–323

Gardea-Torresdey JL, Gomez E, Peralta-Videa JR et al (2003) Alfalfa sprouts: a natural source for the synthesis of silver nanoparticles. Langmuir 19(4):1357–1361

Gengan RM, Anand K, Phulukdaree A, Chuturgoon A (2013) A549 lung cell line activity of biosynthesized silver nanoparticles using *Albizia adianthifolia* leaf. Coll Surf B Biointerfaces 105:87–91

Ghoshal S, Prasad BK, Lakshmi V (1996) Antiamoebic activity of *Piper longum* fruits against *Entamoeba histolytica in vitro* and *in vivo*. J Ethnophorm 50(3):167–170

Glusker JP, Katz A, Bock CW, Rigaku J (1999) Metal ions in biological systems. Glusker Jenny P 6:8–16

Govindaraju K, Tamilselvan S, Kiruthiga V, Singaravelu G (2010) Biogenic silver nanoparticles by *Solanum torvum* and their promising antimicrobial activity. J Biopest 3(1):394–399

Hamburger M, Hostettmann K (1991) Bioactivity in plants: the link between phytochemistry and medicine. Phytochemistry 30(12):3864–3874

Himejima M, Kubo I (1991) Antibacterial agents from the cashew *Anacardium occidentale* (Anacardiaceae) nut shell oil. J Agr Food Chem 39(2):418–421

Holst B, Williamson G (2008) Nutrients and phytochemicals: from bioavailability to bioefficacy beyond antioxidants. Curr Opi Biotech 19(2):73–82

Hu ML (2011) Dietary polyphenols as antioxidants and anticanceragents: more questions than answers. Chang Gung Med J 34(5):449–460

Husen A, Siddiqi KS (2014) Phytosynthesis of nanoparticles: concept, controversy and application. Nanoscale Res Lett 9(1):1–24

Iravani S (2011) Green synthesis of metal nanoparticles using plants. Green Chem 13(10):2638–2650

Jacob SJP, Finub JS, Narayanan A (2012) Synthesis of silver nanoparticles using *Piper longum* leaf extracts and its cytotoxic activity against Hep-2 cell line. Coll Surf B Biointerfaces 91:212–214

Jha AK, Prasad K, Kumar V, Prasad K (2009) Biosynthesis of silver nanoparticles using *Eclipta* leaf. Biotech Prog 25(5):1476–1479

Jiang J, Oberdörster G, Biswas P (2009) Characterization of size, surface charge, and agglomeration state of nanoparticle dispersions for toxicological studies. J Nanopart Res 11(1):77–89

Johnson IT (2007) Phytochemicals and cancer. Proceed Nut Soc 66(02):207–215

Jones SB Jr, Luchsinger AE (1986) Plant systematics. McGraw-Hill Book Co, New York

Kasthuri J, Kathiravan K, Rajendiran N (2009) Phyllanthin assisted biosynthesis of silver and gold nanoparticles: a novel biological approach. J Nanopart Res 11:1075–1085

Katti K, Chanda N, Shukla R et al (2009) Green nanotechnology from cumin phytochemicals: generation of biocompatible gold nanoparticles. Int J Green Nanotech: Biomed 1(1):B39–B52

Khan MA, Khan T, Nadhman A (2016) Applications of plant terpenoids in the synthesis of colloidal silver nanoparticles. Adv Coll Interface Sci 234:132–141

Kim J, Rheem Y, Yoo B, Chong Y et al (2010) Peptide-mediated shape-and size-tunable synthesis of gold nanostructures. Acta Biomater 6(7):2681–2689

Kreuter J (2007) Nanoparticles—a historical perspective. Int J Pharm 331(1):1–10

Kumari M M, Philip D (2013) Facile one-pot synthesis of gold and silver nanocatalysts using edible coconut oil. Spectrochimica Acta Part A Mol Biomol Spect 111:154–160

Li X, Xu H, Chen ZS, Chen G (2011) Biosynthesis of nanoparticles by microorganisms and their applications. J Nanomat 8:1–16

Lisar SYS, Motafakkerazad R, Hossain MM (2012) Water stress in plants: causes, effects and responses, Water Stress, Prof. Ismail Md. Mofizur Rahman (Ed.), ISBN: 978-953-307-963-9, InTech

Lukman AI, Gong B, Marjo CE, Roessner U, Harris AT (2011) Facile synthesis, stabilization, and anti-bacterial performance of discrete Ag nanoparticles using *Medicago sativa* seed exudates. J Coll Interface Sci 353(2):433–444

Makarov VV, Love AJ, Sinitsyna OV et al (2014) "Green" nanotechnologies: synthesis of metal nanoparticles using plants. Acta Nat 6(1):20

Mandal S, Selvakannan PR, Phadtare S, Pasricha R, Sastry M (2002) Synthesis of a stable gold hydrosol by the reduction of chloroaurate ions by the amino acid, aspartic acid. J Chem Sci 114(5):513–520
Martinez MJ, Betancourt J, Alonso-Gonzalez N, Jauregui A (1996) Screening of some Cuban medicinal plants for antimicrobial activity. J Ethnopharmacol 52(3):171–174
Mathur M (2014) Properties of phtyo-reducing agents utilize for production of nano-particles, existing knowledge and gaps. Int J Pure App Biosci 2(2):113–130
Maynard AD, Aitken RJ, Butz T et al (2006) Safe handling of nanotechnology. Nature 444(7117):267–269
McDevitt JT, Schneider DM, Katiyar SK, Edlind TD (1996) Berberine: a candidate for the treatment of diarrhea in AIDS patients, abstr. 175. In: Program and abstracts of the 36th interscience conference on antimicrobial agents and chemotherapy. American Society for Microbiology, Washington, DC
Mittal AK, Chisti Y, Banerjee UC (2013) Synthesis of metallic nanoparticles using plant extracts. Biotechnol Adv 31(2):346–356
Mohanpuria P, Rana NK, Yadav SK (2008) Biosynthesis of nanoparticles: technological concepts and future applications. J Nanopart Res 10(3):507–517
Mubarak AD, Thajuddin N, Jeganathan K, Gunasekaran M (2011) Plant extract mediated synthesis of silver and gold nanoparticles and its antibacterial activity against clinically isolated pathogens. Coll Surf B: Biointerfaces 85(2):360–365
Mukunthan KS, Balaji S (2012) Cashew apple juice (*Anacardium occidentale* L.) speeds up the synthesis of silver nanoparticles. Int J Green Nanotech 4(2):71–79
Murray MT (1998) Quercitin: nature's antihistamine. Better Nutr 60(4):10
Navarro V, Villarreal ML, Rojas G, Lozoya X (1996) Antimicrobial evaluation of some plants used in Mexican traditional medicine for the treatment of infectious diseases. J Ethnopharmacol 53(3):143–147
Nishino H, Satomi Y, Tokuda H, Masuda M (2007) Cancer control by phytochemicals. Curr Pharm Des 13(33):3394–3399
Njagi EC, Huang H, Stafford L et al (2010) Biosynthesis of iron and silver nanoparticles at room temperature using aqueous sorghum bran extracts. Langmuir 27(1):264–271
Ofek I, Goldhar J, Sharon N (1996) Anti-*Escherichia coli* adhesin activity of cranberry and blueberry juices. In: Toward anti-adhesion therapy for microbial diseases. Springer US, Boston, pp 179–183
Omulokoli E, Khan B, Chhabra SC (1997) Antiplasmodial activity of four Kenyan medicinal plants. J Ethnopharmacol 56(2):133–137
Osato JA, Santiago LA, Remo GM, Cuadra MS, Mori A (1993) Antimicrobial and antioxidant activities of unripe papaya. Life Sci 53(17):1383–1389
Park Y, Hong YN, Weyers A, Kim YS, Linhardt RJ (2011) Polysaccharides and phytochemicals: a natural reservoir for the green synthesis of gold and silver nanoparticles. IET Nanobiotech 5(3):69–78
Peteros NP, Uy MM (2010) Antioxidant and cytotoxic activities and phytochemical screening of four Philippine medicinal plants. J Med Plants Res 4(5):407–414
Raghunandan D, Bedre MD, Basavaraja S et al (2010) Rapid biosynthesis of irregular shaped gold nanoparticles from macerated aqueous extracellular dried clove buds (*Syzygium aromaticum*) solution. Coll Surfaces B Biointerfaces 79(1):235–240
Rana BK, Singh UP, Taneja V (1997) Antifungal activity and kinetics of inhibition by essential oil isolated from leaves of *Aegle marmelos*. J Ethnopharmacol 57(1):29–34
Rao SR, Ravishankar GA (2002) Plant cell cultures: chemical factories of secondary metabolites. Biotech Adv 20(2):101–153
Rejeski D, Lekas D (2008) Nanotechnology field observations: scouting the new industrial west. J Clean Prod 16(8):1014–1017
Roy N, Barik A (2010) Green synthesis of silver nanoparticles from the unexploited weed resources. Int J Nanotech Appl 4(2):95–101

Sangwan NS, Farooqi AHA, Shabih F, Sangwan RS (2001) Regulation of essential oil production in plants. Plant Growth Reg 34(1):3–21

Satrija F, Nansen P, Murtini S, He S (1995) Anthelmintic activity of papaya latex against patent *Heligmosomoides polygyrus* infections in mice. J Ethnopharmacol 48(3):161–164

Scheel LD (2016) The biological action of the coumarins. Microb Tox 8:47–66

Sheny DS, Mathew J, Philip D (2012) Synthesis characterization and catalytic action of hexagonal gold nanoparticles using essential oils extracted from *Anacardium occidentale*. Spectrochimica Acta Part A Mol Biomol Spectrosc 97:306–310

Shiv Shankar S, Ahmad A, Pasricha R (2003) Sastry M. J Mater Chem 13:1822–1846

Si S, Mandal TK (2007) Tryptophan-based peptides to synthesize gold and silver nanoparticles: a mechanistic and kinetic study. Chem A Eur J 13(11):3160–3168

Siddiqui IA, Sanna V, Ahmad N, Sechi M, Mukhtar H (2015) Resveratrol nanoformulation for cancer prevention and therapy. Annals NY Acad Sci 1348(1):20–31

Sies H, Stahl W, Sundquist AR (1992) Antioxidant functions of vitamins. Ann NY Acad Sci 669(1):7–20

Silva LP, Reis IG, Bonatto CC (2015) Green synthesis of metal nanoparticles by plants: current trends and challeges. In: Vladimir AB, Elena VB (eds) Green processes for nanotechnology from inorganic to bioinspired nanomaterials. Springer, Cham, pp 259–275

Singh AK, Talat M, Singh DP, Srivastava ON (2010) Biosynthesis of gold and silver nanoparticles by natural precursor clove and their functionalization with amine group. J Nanopar Res 12(5):1667–1675

Smitha SL, Philip D, Gopchandran KG (2009) Green synthesis of gold nanoparticles using *Cinnamomum zeylanicum* leaf broth. Spectrochimica Acta Part A Mol Biomol Spec 74(3):735–739

Stephen A, Seethalakshmi S (2013) Phytochemical synthesis and preliminary characterization of silver nanoparticles using hesperidin. J Nanosci 2013(6):1–6

Tan YN, Lee JY, Wang DI (2010) Uncovering the design rules for peptide synthesis of metal nanoparticles. J Amr Chem Soc 132(16):5677–5686

Vasanthi P, Ganapathy M, Evanjelene VK et al (2014) Phytochemical screening and antioxidant activity of extracts of the leaf and bark of *Albizzia lebbeck* (Benth). Acad J Med Plant 2:026–031

Verpoorte R, Van der Heijden R, Ten Hoopen HJG, Memelink J (1999) Metabolic engineering of plant secondary metabolite pathways for the production of fine chemicals. Biotechnol Lett 21(6):467–479

Vijayakumar M, Priya K, Nancy FT, Noorlidah A, Ahmed ABA (2013) Biosynthesis, characterisation and anti-bacterial effect of plant-mediated silver nanoparticles using *Artemisia nilagirica*. Ind Crop Prod 41:235–240

Vilas V, Philip D, Mathew J (2014) Catalytically and biologically active silver nanoparticles synthesized using essential oil. Spectrochimica Acta Part A Mol Biomol Spect 132:743–750

Wan J, Wilcock A, Coventry MJ (1998) The effect of essential oils of basil on the growth of *Aeromonas hydrophila* and *Pseudomonas fluorescens*. J Appl Microbiol 84(2):152–158

Wang WX, Vinocur B, Shoseyov O, Altman A (2000) Biotechnology of plant osmotic stress tolerance physiological and molecular considerations. In IV international symposium on in vitro culture and horticultural breeding 560:285–292

Wild R (1994) The complete book of natural and medicinal cures. Rodale Press, Inc, Emmaus, pp 50–56

Wiley B, Sun Y, Mayers B, Xia Y (2005) Shape controlled synthesis of metal nanostructures: the case of silver. Chem A Eur J 11(2):454–463

Zamiri R, Azmi BZ, Sadrolhosseini AR, Ahangar HA, Zaidan AW, Mahdi MA (2011) Preparation of silver nanoparticles in virgin coconut oil using laser ablation. Int J Nanomedicine 6:71

Zayed MF, Eisa WH, Shabaka AA (2012) Spectrochim. Acta A Mol Biomol Spectrosc 98:423–428

Zwenger S (2008) Plant terpenoids: applications and future potentials. Biotechnol Mol Biol Rev 3(1):1–7

Chapter 3
Classification of Green Nanoparticles

Beenish Zia Butt

3.1 Introduction

Nanotechnology is an interdisciplinary field. It is being considered as an innovative and promising field of twenty-first century which involves the synthesis of materials at nano scale level (Kumar et al. 2011). Nano science is an impending area of concern in the field of material sciences (Berry and Curtis 2003). The word "Nano" has Greek origin meaning extremely small. Nanoparticles are the collections of atoms in the size range of 1–100 nm (Gong et al. 2007). Nanotechnology has expected capability to fabricate things from the base, using systems and devices that are being made to construct elevated performance inventions. The field of nanobiotechnology connects two different areas of science, biotechnology and material science. It involves 12 general principles of green chemistry that comprises of these objectives (to avoid devastate, use of unprocessed stuff, reduce the use of hazardous compounds, high pressure and temperature (Pal et al. 2011; Singh et al. 2014; Gardea-Torresdey et al. 2003).

3.2 Green Nanotechnology

In 2003 the earliest account on the green synthesis of silver nanoparticles using alfalfa was reported which was the first step towards green nanotechnology (Gardea-Torresdey et al. 2003). Green synthesis approaches using different biological methods including bacteria, yeast, algae and plant extracts have been established for nanoparticles fabrication (Kaviya et al. 2011). The application of chemistry rules to lessen or abolish the harmful chemicals has caused the considerable decrease in the

B. Z. Butt (✉)
Department of Botany, University of Punjab, Lahore, Pakistan

S. Javad, A. Butt (eds.), *Nanobotany*, https://doi.org/10.1007/978-3-319-77119-9_3

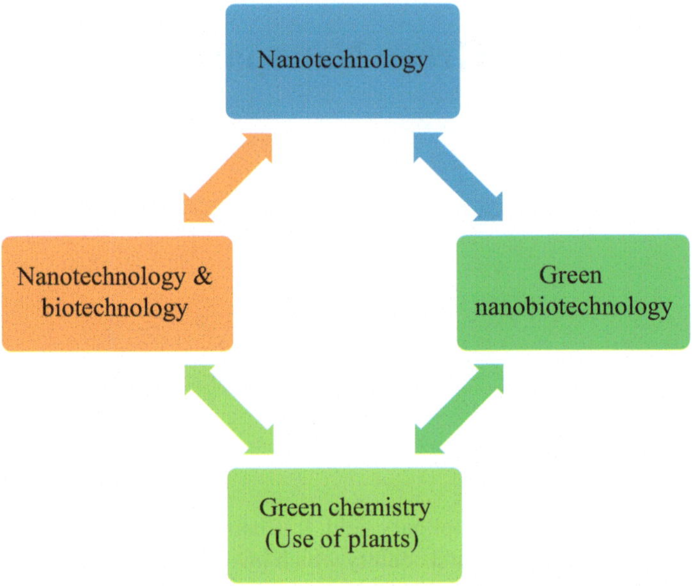

Fig. 3.1 Interrelated fields of chemistry, nanotechnology and biotechnology

quantity of toxic compounds into the environment. Green chemistry is the use of chemistry assisted procedures which anticipates pollution (Melchert et al. 2012). The use of plants and different plant resources for the fabrication of novel metal nano sized materials is referred as "green nanotechnology". Green chemistry is the exploitation of standard principles that decreases or eradicates the utilization and production of harmful compounds in the creation and the relevance of chemical products (McKenzie and Hutchison 2004) (Fig. 3.1).

It is faster and reliable method used to scale up the production process of commercially relevant NPs with less or no toxicity. Therefore, plants are used for NPs synthesis as they actively uptake and reduce metal ions in bioremediation and thereby can form complex metal NPs (Singh et al. 2014).

Nowdays, biogenic fabrication of metallic nanoparticles using plants is gaining much importance as this route of synthesis is easiest, simple, time saving and rapid for the formation of nanoparticles with unique shapes (Pandey et al. 2012).

3.3 Plants and Green Nanotechnology

Exploitation of plant extracts in comparison to entire plant part for the synthesis of nanoparticles has shown significant outcomes. In the present century, green synthesis of nanoparticles using plant extracts is gaining much attention due to the growing need of nanoparticles (Daisy and Saipriya 2012; Lee et al. 2011). As the plants contain numerous phytocompounds of biological importance, which are used as therapeutic agents in conventional medications system. Several photochemical like phenols and alkaloids are identified to evaluate the synthesis and stabilization of NPs. Plants rich in genetic variability with diverse range of biological molecules and tremendous reducing agents include phenolic compounds, sterols and alkaloids, metal ions can be reduced into NPs within a single step (Ravishankar and Jamuna 2011).

Plants have the potential to reduce metal ion as evident from the fact that they can grow in soils with high concentration of lethal metals and this property is known as hyper accumulative potential of plants (Kulkarni and Muddapur 2014; Kale et al. 2013). For the synthesis of nanoparticles plants and plant resources are the most attractive biological means. Plants with genetic diversity have various remarkable biological molecules in the form of secondary metabolites that act as reducing and capping agents in nanoparticles synthesis. As the green synthesis of nanoparticles can be performed at room temperature so it requires no temperature, pressure and other complex constraints. Moreover, the green synthesis is easy, simple, cost effective and eco-friendly approach as it reduces the use and generation of noxious chemicals. Plant biomolecules (alkaloids, phenolics and sterols) act as exceptional reducing and capping agents in nanoparticles formation. In green route synthesis of nanoparticles whole plant or plant part can also be used in addition to plant extract that is considered a major advantage of this method (Park et al. 2011).

Biologically potent biochemicals of plants including phenolic acids, sugars, alkaloids, proteins, terpenoids and polyphenols are considered to have a vital function in the reduction of metal ions (Marshall et al. 2007) (Fig. 3.2).

Plants extracts are supposed to be the excellent biological agents that reduce and also stabilize the synthesized nanoparticles whereas biochemical nature of plant extracts greatly influences the shape and size of resulted nanoparticles (Mukunthan and Balaji 2012). This is considered due to the varied concentration of phytochemicals in different plants or plant parts (Lee et al. 2011). For the biosynthesis of nanoparticles, at room temperature plant extract is added into precursor (metal salt solution). The biochemical reduction of metal ions completes within a short period of time that can be indicated by observing the change in solution color. On reduction metals are switched to zero valent condition. This indicates the synthesis of nanoparticles, biosynthesis of silver, iron, gold and many other metals which have been described (Safaepour et al. 2009).

3.4 Green Nanoparticles

Nanoparticles are mainly divided into two large groups; one group contains carbon based nanoparticles that are organic nanoparticles while second group covers magnetic nanoparticles. Novel metallic nanoparticles besides this are metal oxide nanoparticles which are also known as semi conductor nanoparticles. Metal nanoparticles are the essential excellent agents used in biomedical. Some of the

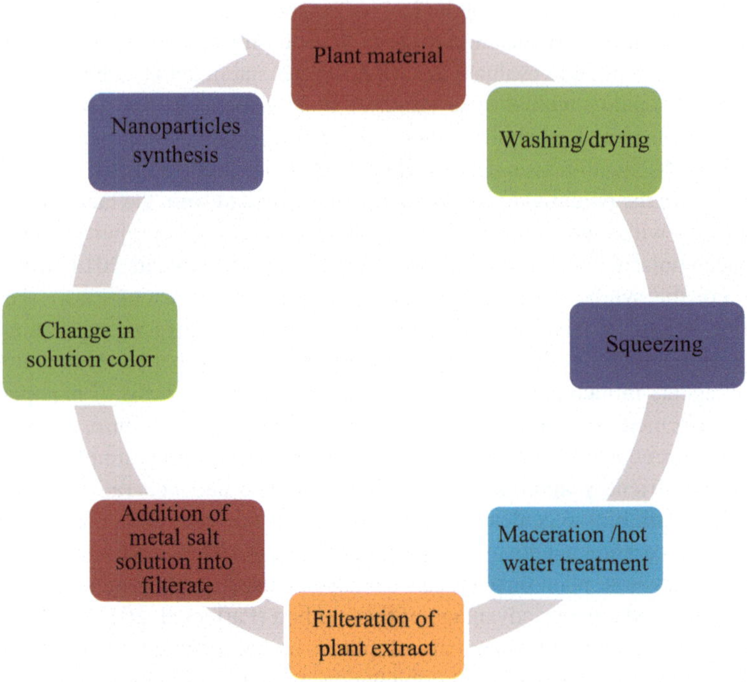

Fig. 3.2 Mechanism of green nanoparticles synthesis

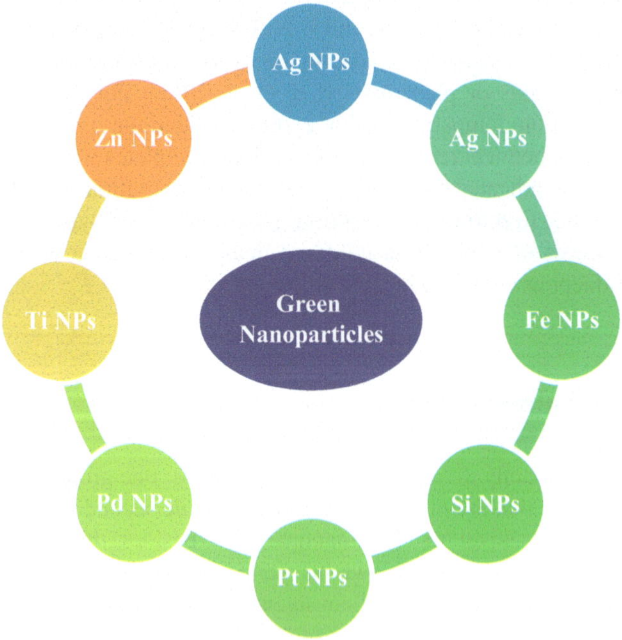

Fig. 3.3 Different types of green nanoparticles

consistently used nanoparticles are gold, silver, iron, aluminum, copper, palladium, zinc and titanium (Gonzalez and Noguez 2007) (Fig. 3.3).

3.5 Nanoparticles Composition

Nanoparticles can be characterized on the basis of their composition; these can be made up of only a particular component or may be composed of more than one constituent's materials. Naturally occurring nanoparticles are mostly a grouped collection of different materials though the nanoparticles with a distinct composition can be fabricated using various techniques (Janka and Maldarelli 2004).

Plant based nanoparticles have a great range of morphology and diameter. Two most important biomolecules that contribute in the shape variations of nanoparticles are peptides and proteins and they also influence the yield of nanoparticles. Generally the characteristics of nanoparticles are highly affected by the nature and surface morphology of nanoparticles (Lynch et al. 2006).

The diversity in concentration and composition of bioactive molecules within different plants and their consequent mixing with metal salt solution is said to be the major cause that is responsible for the variations in nanoparticles morphology (Lee et al. 2011).

3.6 Basic Approaches for Nanoparticles Synthesis

Basically two approaches are used to synthesize nanoparticles one is top down and the other is bottom up method (Sepeur 2008).

1. **Top-down synthesis**

In this method nanoparticles are synthesized from an appropriate initiating material that produces nanoparticles by reducing size (Meyers et al. 2006). Reduction in size can be done using other procedures. This method causes flaws in structure of product which is a main drawback because various distinct characters of nanoparticles depend on their surface structure (Thakkar et al. 2010).

2. **Bottom up synthesis**

In this method nanoparticles are synthesized by the joining of small units (atoms), molecules and other smaller things. In this process, first building units of nanoparticles are synthesized which then grouped and form the complete nanoparticle (Thakkar et al. 2010; Mukherjee et al. 2001). This method of nanoparticles synthesis generally depends on biological and chemical processes. From the biological synthesis of nanoparticles, microorganisms based nanoparticles have been greatly reported (Kaler et al. 2011) (Fig. 3.4).

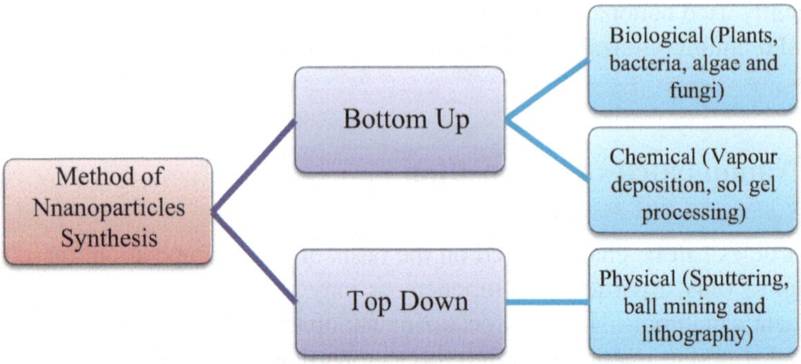

Fig. 3.4 Methods of nanoparticles synthesis

3.7 Methods for the Green Synthesis of Nanoparticles

There are few methods for the synthesis of green nanoparticles (Bhosale et al. 2014).

1. **Polysaccharide method**

   In this method polysaccharides are used as reducing and capping agents that reduce the metal ions into nanoparticles. In this method water is used as solvent as it acts as benign solvent universally (Raveendran et al. 2003).

2. **Tollens method**

   In this method saccharides are used with ammonia to reduce metals ions into nanoparticles. Nanoparticles synthesized from this method have defined diameters.

3. **Polyoxometalates method**

   Polyoxometalates are used in this method that are water soluble and undergo a series of reduction oxidation reactions and reduce the metal ions into nanoparticles and act as stabilizing agent.

4. **Biological method**

   In this method biological materials are used to synthesize nanoparticles, these may be microbes (bacteria and fungi) or plant materials that act as metal ion reducers.

The process of green synthesis of nanoparticles using plant extracts and plants covers three main steps.

(1) In the first step the bioreduction of metal ions using plant extracts occurs besides the nucleation of reduced metal atoms.

(2) In second step the minute adjoining nanoparticles instinctively unite into clusters of large sized particles that leads to the synthesis of nanoparticles which is followed by the stabilization of nanoparticles.
(3) It's the third and last stage that determines final morphology of synthesized nanoparticles (Si and Mandal 2011).

3.7.1 Chemical Synthesis of Nanoparticles

Few methods that are employed for the chemical synthesis of nanoparticles are as follows.

1. **Solvent evaporation method**

This method involves the mixing of polymer in an organic solvent (chloroform, dichloromethane or ethyl acetate), that also act as a solvent for dissolving the hydrophobic drug. The solution of drug and polymer is then emulsified in a solution containing a surfactant or softening agent to form oil in water (o/w) mixture. Organic solvent is evaporated either by continuous moving or reducing the pressure after the formation of constant suspension. Particle size of the synthesized nanoparticles is found to be affected by the concentration and nature of stabilizer, homogenizer speed and polymer concentration. In order to synthesize small sized particles, often a high-speed homogenizer may be used (Li et al. 2001).

2. **Solvent diffusion method**

It is an advanced form of solvent evaporation method. In this method, the water soluble solvent along with a small amount of the water insoluble organic solvent is used. Due to the impulsive dispersal of solvents an interfacial instability is formed between the two phases leading to the synthesis of small sized particles. Particles with smaller size can be synthesized as the concentration of water miscible solvent increases. Both solvent diffusion and evaporation methods can be employed for the formation of hydrophilic and hydrophobic drugs (Zhang et al. 2001).

3. **Polymerization method**

In this method, monomers are assembled to form nanoparticle in an aqueous solution. Drug is added either by being mixing in the polymerization solution or by adsorption into the nanoparticles. The formed nanoparticle mixture is then purified to eliminate different surfactants and stabilizers used for polymerization by ultracentrifugation (Boudad et al. 2001).

3.8 Types of Nanoparticles

1. **Organic nanoparticles**

These are solid particles made up of organic compounds ranging in diameter from 10 nm −1 μm. There are four types of organic nanoparticles which are as follows.

(i) **Nanogels** are organic nanoparticles formed of hydrogel (network of hydyrphilic crosslinkages). These nanogels are mostly made up of synthetic polymers or bi polymers that are associated to each other physically or chemically. Nanogels range from 10–100 nm in size. The small unfilled space or pore in nanogels can be filled through tiny molecules and their several properties can be controlled (Bencherif et al. 2009; Kabanov and Vinogradov 2009; Li et al. 2007; Hayashi et al. 2004).

(ii) **Liposomes** are organic nanoparticles that are circular or spherical in shape contain one or more lipid bilayers in them. These liposomes are efficiently used in pharmaceutical drugs for drug delivery. Liposomes are also made by disturbing the biological membranes. These are mostly made up of phospholipid, including egg phosphatidylethanolamine, phosphatidylcholine. In liposomes, surface ligands are also used for interactions to an unhealthy cell or tissue (Nath and Banerjee 2013).

(iii) **Micelles** are the collections of surface molecules dispersed in a liquid colloid. Micelles are mostly spherical in morphology. A characteristic micelle in aqueous solution forms a collection with the hydrophilic "head" regions when comes in contact with adjacent solvent, seizing the hydrophobic single-tail part in the micelle middle. This happens because of the assemblage of single lipid tail in a bilayer. The impenetrability fills all the space within the bilayer and accepting the region per head group pressured on the molecule due to the hydration of head that directs the micelle formation. This is simply known as normal micelle. In inverse micelles the head groups are found at the centre while the tails extending out (Slomkowski et al. 2011).

(iv) **Dendrimers** Organic nanoparticles with frequently branching are known as dendrimers. Dendrimers name comes from Dendron a Greek word meaning tree. These are also known as arborols. It has a proper spherical symmetry. Dendrons commonly have a point called core or focal point. The first dendrimers were made by Fritz Vögtle in 1978 (Nanjwade et al. 2009; Buhleier et al. 1978).

2. **Inorganic Nanoparticles**

There are four types of inorganic NPs (Srinivasan et al. 2015) that are described as follows.

(i) **Superparamagnetic iron oxide** particles are inorganic nanoparticles that have been employed in medical sciences to sense the incidence of cancers. Surface alterations to the molecules including layering with surfactants and biological

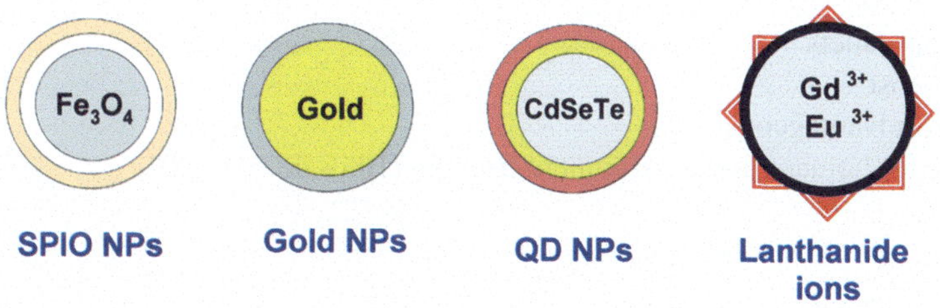

Fig. 3.5 Inorganic NPs Four important inorganic nanomaterials used in the construction of multi-functional NPs are SPIO, gold, QD and lanthanide ions

joining of the amine and carboxyl groups make these acquiescent (Kievit and Zhang 2011).

(ii) **Quantum dots** are extremely small nanoparticles, as their optical and electronic properties are unique than the larger particles. This process is common in semiconductors. QD's are a fundamental subject in nanotechnology. These are also called as artificial atoms. **Quantum dots** are inorganic types of nanoparticles that are in between semiconductors and distinct molecules. The size and shape of QD greatly influences their optical and electric properties. The QD with a bigger size produces radiations with longer wavelength that results in release of the colors. Radiations of shorter wavelength are emitted by the small sized QDs. Due to their great interest QD have variable properties. Furthermore, small size of QDs permits them to be suspended in solution. These techniques result in less-costly and time saving process of semiconductor formation (Ashoori 1996; Murray et al. 2000).

(iii) **Gold NPs** are a class of inorganic nanoparticles that show electrical and optical characteristics. These are also used in the monitoring of cure response and tumor activity (Fig. 3.5).

(iv) **Lanthanide ions** are a class of inorganic nanoparticles that produce fluorescence and permit the transfer or change of low light energy (near infra red) to high visible light energy. The effect of conversion that occurs due to the energy transfer has been used in cancer detection and in many other therapies (Coll 2011; Cheng et al. 2012)

3.9 Benefits of Green Synthesis Over the Chemical Synthesis

There are few advantages of green synthesis that are as follows (Bhosale et al. 2014).

1. Energy efficient
2. Economical

3. Less waste
4. Safe products
5. Low cost
6. More advantageous
7. Save for humans and environment friendly (Bhosale et al. 2014).

3.10 Gold Nanoparticles

Gold is one of the novels and expensive one among all the metals. From last 50 years gold is exploited as therapeutic agent. Its nanoparticles range in size from 100–1000 nm and have been considered as most significant nanoparticles as they are biologically compatible and non toxic. The diameter of gold nanoparticles can be managed at the time of nanoparticles formation using special functional groups. Gold nanoparticles can be used in biomedical as chemotherapeutic and diagnostic agents in cancer treatment (Cai and Chen 2007).

Gold nanoparticles exist in different forms such as gold nanorods, gold nanoshell, gold nanocages and gold nanosphere. These have wide range relevancies in drug release, as biosensors, in protein and gene transformation (Li et al. 2006). Gold nanoparticles are found in a variety of diameters from 2–100 nm, but it is seen that nanoparticles upto 50 nm size demonstrate very significant activity during cellular uptake (El-Sayed et al. 2006). Gold nanoparticles are bioconsistent, chemically stable due to gold sulphur bonds. These are multifaceted due to their linkages with thiol group (Thakor et al. 2011; Connor et al. 2005). These have appeared as capable transforming agents for insulin, proteins and nucleic acid. Their distinctive characteristics including least toxicity, small size, large surface area and versatility make them highly applicable in different fields and have transformed the field of biomedical sciences (Tomar and Garg 2013).

Biosynthesis of gold nanoparticles through plants is most valuable method. In this method high yield of nanoparticles can be obtained in which plant extracts act as bio-reducing chemicals that also stabilize the resultant nanoparticles. During past few decades, the exploitation of plant as reducing agents has gained substantial interest because of its simplicity (Chah et al. 2005). Gold nanoparticles have many distinctive characteristics in comparison to bulk gold that is in solid form and yellow whereas the nanoparticles solution is red in color. Gold nanoparticles with various morphological structures (nanospheres, nanobelts and nanorods) can be synthesized (Mittal et al. 2013). The diameter and morphology of gold nanoparticles greatly affect their physiochemical properties (Ganeshkumar et al. 2012).

Green synthesis of gold nanoparticles using plants is receiving much attention as its metal salt can be easily reduced and nanoparticles have potent bactericidal activity. This method for the synthesis of nanoparticles at commercial scale emerges as appropriate one as it is simple, easy to conduct, low cost and secure for clinical research. Green synthesis of gold nanoparticles from two medicinal plants (*Malva crispa* and *Cucurbita pepo*) was also reported, synthesized nanoparticles showed excellent bactericidal activity against food contaminating microbes (Chandran et al. 2014). Biosynthesis of gold nanoparticles using *Acalypha indica* as reducing agent was reported and nanoparticles with 20–30 nm size were produced (Krishnaraj et al. 2014).

Gold nanoparticles were also fabricated using *Cymbopogon citrates* leaf extract worked as bioreducing and capping material. Both the synthesized nanoparticles and leaf extract were assed against malaria and dengue vector (Murugan et al. 2015). *Zingiber officinale* was also studied for the synthesis of gold nanoparticles which were found to be very small sized nanoparticles (10 nm) (Singh et al. 2011). *Abelmoschus esculentus* seed were used to synthesize gold nanoparticles and synthesized nanoparticles were tested against different fungal species where significant results were observed. Therefore the biosynthesized nanoparticles have an immense importance in the synthesis of antifungal drugs (Jayaseelan et al. 2013). The synthesis of stable gold nanoparticles using *Pelargonium graveolen was* carried out and nanoparticles with different sizes were obtained (Shankar et al. 2003).

Leaf extract mediated synthesis of gold nanoparticles using *Nepenthes Khasiana* was conducted and the synthesized nanoparticles were also characterized through tunnelling electron microscopy (TEM) and scanning electron microscopy (SEM) results revealed the spherical and triangular morphology of nanoparticles with 50–100 nm diameter (Bhau et al. 2015). Aqueous leaf extract of *Cassia auriculata* and coriander was reported for the synthesis of gold nanoparticles (Kumar et al. 2011). Gold nanoparticles for the use in cancer therapy were synthesized from the peel extract of *Punica granutum* (Ganeshkumar et al. 2013). Gold nanoparticles with different size and shapes were fabricated using Magnolia, Hibiscus and Persimmon leaf extracts (Philip 2010; Song et al. 2009). Green synthesis of gold nanoparticles was carried out using various plants and plant extracts including *Basella alba, Ziziphus mouritiana, Murraya koenigii, Aloe vera, Coriandrum sativum, Allium sativum, Hibiscus rosasynesus, Zingiber officinale, Datura metal, Citrus limonium, Azadirachta indica, Mangifera indica, Mentha arvensis* and many other plants have been used for gold nanoparticles (Sumit and Nayak 2012).

3.11 Silver Nanoparticles

Silver is an essential metal of earth, highly recognized one from past times because of its therapeutic significance as it has additive characteristics. It is a proficient anti-bacterial agent than other metals because of its exceptionally high surface area that offers improved contact with microbes. Silver nanoparticles are the novelest metallic nanoparticles which have been researched widely because of their numerous applications in biology (Chu et al. 1988). Silver is a harmless, secure inorganic antimicrobial agent employed from ancient times and has the significant inhibitory potential against pathogenic microbes (Jeong et al. 2005).

Silver nanoparticles are the particles of silver having a diameter up to 100 nm and they have antibacterial action in preparation of silver based products (Prabhu and Poulose 2012). Silver nanoparticles showed wide range of antibacterial and antifungal activities which has made them enormously accepted (Tran and Le 2013). Silver nanoparticles with varied morphology including globular, triangular and rod shaped have been synthesized using different methods (Vadlapudi et al. 2014).

The use of silver nano materials in nanotechnology seems to be very potent. It is studied that silver nanosized particles are safe for living cells therefore these do not aggravate pathogens resistance. It is assumed that permeability of cell wall and respiration is greatly influenced as silver nanoparticles get attached to cell wall (Singh et al. 2008). Silver nanoparticles have been found to be most beneficial as compared to other nanoparticles as they are very effective against micro-organisms and safe for human beings (Jeong et al. 2005).

Traditionally, silver is acknowledged for its bactericidal properties and hence it is frequently exploited in numerous pharmaceutical products that are used against microbes (Parak et al. 2003; Kumar and Yadav 2009). Silver is well known for its strong antiviral, antibacterial and antifungal activities while according to modern research, the application of silver as poisonous substance (biocide) specifically in nano dimensions form has practiced a remarkable recovery. Silver at nano sized level (nanosilver) has been recently utilized in a variety of products. The main cause in the current boost in nano-silver based products is its excellent antibacterial potential activity (Sharma et al. 2009). The synthesis of silver nanoparticles can be carried out through microorganisms and plant extracts (Prabhu and Poulose 2012). The chief advantage of green synthesis approach for the formation of nanoparticles is to develop highly effective particles for biomedical use as this method of nanoparticles fabrication does not involve synthetic harmful chemicals (Geetha et al. 2012).

Amongst the biological methods for the fabrication of nanoparticles, use of plants proved to be most advantageous that supply a favorable and speedy approach for the formation of silver nanoparticles in relation to other traditional pathways (Edison and Sethuraman 2012). At present one of the main feature of the nanobiotechnology is the progress in green synthesis route for the fabrication of silver nanoparticles. Synthesis of nanoparticles using plants is beneficial than the other methods and cost effective. Plants are most significant substitute for nanomaterial formation as they produce natural save biochemicals that act as excellent cappig and reducing agents. As silver nanoparticles have a variety of applications in various fields where they are exploited as food and fabric preservatives in textile industry and food industry to eliminate microbes (Elumalai et al. 2010). Antibacterial potential of silver has a main role in foodstuff packaging processes, while the use of silver nanomaterials in microbial resistant paints is found to be safe and eco-friendly (Kumar et al. 2008). It is observed that silver nanoparticles demonstrate capacity to cure wounds without leaving marks and also widely used in cosmetic industry (Tian et al. 2007).

Cucumis Sativus fruit extract was used for the biosynthesis of silver nanoparticles. Silver nitrate solution was used for the synthesis of silver nanoparticles. The synthesis of AgNPs was investigated using UV visible spectroscopy, transmission electron microscopy (TEM) and x-ray diffraction (XRD). Antibacterial activity of the synthesized AgNPs was checked against *Klebsiella pneumonia, Escherichia coli* and *Staphylococcus aureus.* Significant results were obtained (Roya et al. 2015).

Aqueous extract of the *Basella alba* was used for the biosynthesis of AgNPs. A change in solution color was observed on the bioreduction of metal ions. The structural properties of the resulted AgNPs were investigated through UV visible spectroscopy, SEM, TEM and FTIR. Spherical morphology of the synthesized nanoparticles was observed with a diameter 22–25 nm (Kl et al. 2014).

Green synthesis of silver nanoparticles was carried out using mulberry leaves extract. Silver nitrate was used as a precursor while leaves extracts were used as reducing agent. The biosynthesized silver nanoparticles were further characterized using X-ray diffraction (XRD) and scanning electron microscopy (SEM). Useful results were seen when bactericidal effect of AgNPs was assessed against *Shigella* spp. and *Staphylococcus aureus* (Awwad and Salem 2012) (Table 3.1).

3.12 Copper Nanoparticles

Copper is an important microelement that acts as structural component in many vital metabolic processes and necessary for plant growth and metabolism (Passam et al. 2007). It is quite economical metal and pocesses significant bactericidal potential than other valuable and expensive metals such as silver and gold (Honary et al. 2012). In recent days the use of nanoparticles has increased because of their extensive use in biology, it is a known fact that inorganic nanoparticles have remarkable antibacterial activity and these antimicrobial nanoparticles have a useful application in pharmaceutical preparations (Pillai et al. 2012). Synthesis of copper nanoparticles from biological material is an active area of investigation in nanobiotechnology. This is rising as a promising field of science for the synthesis of desired nanoparticles with unique properties.

Currently, the biosynthesis of copper nanoparticles using plants has been reported and it is an important approach in nanobiotechnology (Honary et al. 2012). The biosynthesis of copper nanoparticles has been reported using numerous plants. Magnolia leaf extract was used for the biofabrication of copper nanoparticles, stable copper nanoparticles with 40–100 nm sizes were synthesized which showed significant bactericidal action against pathogenic bacterial species. Copper nanopaericles were synthesized from the extract of *Syzygium aromaticum*, nanoparicles with a mean particle size of 40 nm and a spherical to granular morphology were obtained. Copper nanoparticles of globular shapes were also synthesized using *Syzygium aromaticum* and *Euphorbia nivulia* (Subhankari Nayak 2013; Valodkar et al. 2011). The biosynthesis of copper nanoparticles was done through the leaf extract of *Ocimum sanctum*. Stable copper nanoparticles due to the main bioactive phytochemicals of tulsi were fabricated and confirmed using different characterization techniques (Kulkarni and Kulkarni 2013).

Leaves of *Ginkgo biloba* were used for the synthesis of copper nanoparticles. Plant extract acted as bioreducing agent that also stabilized the synthesized copper nanoparticles. Resultant nanoparticles were examined and characterized through various analytical techniques (Nasrollahzadeh and Sajadi 2015). *Medicago sativa*

Table 3.1 Use of various plants in the biosynthesis of gold and silver nanoparticles (Mishra et al. 2014)

	Plants	Size (nm)	Shape	Nanoparticles
1	*Allium cepa (onion)*	54	Spherical	Ag
2	*Aloe vera*	25	Spherical	Ag
3	*Azadrachta indica,*	59	Spherical	Ag
4	*Black tea*	25	–	Ag, Au
5	*Brassica campertris*	25	Spherical	Ag
6	*Caesalpinia pulcherrima*	50	Spherical	Au
7	*Capsicum annuum*	–	Spherical	Ag
8	*Cassia auriculata*	–	Sphereical	Ag
9	*Cinnamommum camphora*	80	Triangular	Ag
10	*Cinnamon zeylanicum*	40	Rod shaped	Ag
11	*Cycas*	3.29	–	Ag
12	*Cymbopogon flexuosus*	200	Hexagonal	Ag
13	*Eclipta*	55	Spherical	Ag
14	*Elaeognus latifolia*	50	Spherical	Ag
15	*Emblica officinalis*	20	Spherical	Ag, Au
16	*Eucalyptus camaldulensis*	20		Ag, Au
17	*Euphorbia hitra*	50	Spherical	Ag
18	*Euphorbia prostrats*	80	Rod shapes	Ag
19	*Ficusbenghalensis*	44	Spherical	Ag
20	*Gliricidia sepium*	27	Spherical	Ag
21	*Helianthus annus*	–	Spherical	Ag
22	*Jatropha curcas*	50	–	Ag
23	*Loquat*	18	Spherical	Ag
24	*Medicago sativa*	200	Irregular	Au
25	*Memecylon umbellatum*	20	Spherical	Ag, Au
26	*Mulberry*	40	Spherical	Ag
27	*Murraya koenigii*	20	Hexagonal	Ag
28	*Nerium oleander*	67	–	Ag
29	*Ocimum sanctum*	20	Spherical	Ag
30	*Ocimum basillicum*	89	Spherical	Ag
31	*Oryza sativa,*	25	Spherical	Ag
32	*Pelargomum graveolens*	40	–	Ag
33	*Punica granatum*	400	Spherical	Ag
34	*Putranjiva roxburghi*	38	Spherical	Au
35	*Rosa domascena*	21	Spherical	Ag
36	*Saururus chinenis*	54	Spherical	Ag
37	*Sesbania*	20	Spherical	Au
38	*Shorea tumbuggaia*	403	–	Ag
39	*Solanum xanthocarpum*	406	Spherical	Ag
40	*Tagetes erecta*	10	Spherical	Ag
41	*Tamarindus aestivum*	100	Triangular	Au

(continued)

Table 3.1 (continued)

	Plants	Size (nm)	Shape	Nanoparticles
42	*Trianthema decandra*	50	Hexagonal	Ag
43	*Tridax procumbens*	55	–	Ag
44	*Vitex negundo*	30	Spherical	Ag
45	*Withania somnifera*	40	Spherical	Ag

and *Lantana camara* leaf extracts were exploited for their potential to synthesize copper nanoparticles, (20 nm) average diameter of *L. camara* synthesized nanoparticles was observed (Majumder 2012). Copper oxide nanoparticles were fabricated using green synthesis approach from *Gloriosa superb*. Globular shaped nanoparticles with an average size of 8 nm were produced. Synthesized CuO nanoparticles were tested against some pathogenic bacterial species that showed effective results (Naika et al. 2015).

3.13 Iron Nanoparticles

Iron is a vital fourth copious ubiquitous transition metal of the earth crust. Thus it is absurd that iron as nanoparticle has been ignored because of its oxides and new metals. Although, nano sized iron has immense concern due to its different applications (Huber 2005).

Nanosized iron particles are a subject of great significance. These nanoparticles have various applications in different areas of science (Berry and Curtis 2003; Koo et al. 2012). In the current century, the main concern is the fabrication of eco-friendly, inexpensive and non toxic iron nanoparticles. Green fabrication of FeNPs through plants is incredible method as it is free from the requirements of harmful chemicals, temperature and pressure (Pattanayak and Nayak 2013). Unluckily, iron nanoparticles have been unobserved to some extent, but also understandable. It is difficult to explore the iron nanoparticles as they are extremely reactive. However, nano sized iron has vast interest due to possible catalytic and magnetic application (Huber 2005).

Biosynthesis of iron nanoparticles was achieved using lemon leaf extract. The biological assays of biosynthesized nanoparticles were assessed against four fungal and four bacterial species. Iron nanoparticles demonstrated excellent bactericidal activity against *S. aureus and B. subtilis* as compared to standard antibacterial drug and it showed nearly similar inhibition activity against two fungal species. Nanoparticles were found to be spherical in shapes. Size range of nanoparticles was determined by TEM and it was found to 76 nm (Jayandran et al. 2015). Aqueous extracts of ten selected plant parts (clove buds, mango leaves, black tea, coffee seeds, green tea leaves, cumin seeds, rose leaves, thymol seeds, origano leaves, curry leaves) were used to synthesize iron nanoparticles. The successful formation of iron nanoparticles was established through SEM and UV-visible spectrophotometer (Pattanayak and Nayak 2013).

Six different plants were used for the fabrication of iron nanoparticles from south gujarat regions. The resulted nanoparticles were analyzed and characterized by using different techniques. The size of Fe NPs was examined in the range of 13–342 nm (Shah et al. 2014). Iron nanoparticles were fabricated from *Dodonaea viscose, Gardenia jasminoides* and *Lawsonia inermis*. The biosynthesized NPs were observed as globular in shape and the maximum size was calculated as 27 nm. Nanoparticles were found to be effective against all tested bacterial species (Daniel et al. 2013; Naseem and Farrukh 2015). Bio-fabrication of iron oxide nanoparticles was carried out using *Caricaya papaya* leaf extract. FTIR study revealed the most probable groups (CH, carbonyl and OH) that were responsible for the nanoparticles synthesis. Nanoparticles of irregular shapes with 33 nm were observed (Latha and Gowri 2014).

Biogenic synthesis of iron and iron oxide nanoparticles was done using aqueous leaf extract of *Terminalia chebula* and *Eucalyptus globules* (Balamurugan et al. 2014). Zero valent iron nanoparticles were fabricated from mango leaf using simple and fast green strategy. The chinonin and mangiferin are the most potent photochemicals of mango. It is one of the most essential flavonoid and also shows considerable antioxidant activity. The size of FeNPs was measured mainly between 50–100 nm (Dhuper et al. 2012). Iron oxide nanoparticles were fabricated through green method. In this investigation iron oxide nanoparticles were synthesized from the leaves extract of *Tridax procumbens*. Biosynthesized nanoparticles were formed due to the presence of carbohydrates in plant extract (Senthil and Ramesh 2017).

Magnolia champaca, Datura metel and *Chenopodiastrum murale* leaves extracts were used separately for the formation of iron NPs. Iron chloride was prepared for the biosynthesis of nanoparticles. The maximum size of NPs was measured between 20–150 nm. Biofabricated nanoparticles were found to be round in morphology. Biogenic formation of iron nanoparticles from *Ocimum sanctum* leaf extract was achieved. Morphology and size of prepared particles was determined through transmission electron microscopy and scanning electron microscopy. Maximum particle size of FeO nanoparticles was found to be 47 nm. This study concludes that biogenic fabrication of FeO-NPs through tulsi extract can be cost effective and successful process to synthesize nanoparticles (Balamurugan et al. 2014).

Metal nanoparticles as bactericidal agents are gaining much attention while the little is known about iron nanoparticles. Iron nanoparticles (zero valent and oxides) showed antibacterial activity, zero valent iron nanoparticles proved to be very effective in killing pathogenic bacteria (*E. coli*). Iron oxide produces free radicals that are responsible for the bactericidal action of iron. Iron oxide are considered most important due to their multivalent states and distinctive properties of nano materials, these are quite inert and safe (Cornell and Schwertmann 2003). Iron oxide nanoparticles have involved serious research concern because of their numerous significant submissions including bactericidal activity, cancer treatment, drug release, magnetic resonance imaging (MRI) and wastewater management (Mody et al. 2010).

3.14 Palladium Nanoparticles

Palladium is a rare and lustrous silvery-white metal. Biosynthesis of palladium nanoparticles was reported using plants and plant extracts of cinnamon bark, *Annona squamosa* (Sathishkumar et al. 2009). Green synthesis of palladium nanoparticles was also studied from the leaf extract of magnolia and soybean glycine max (Lee et al. 2011)

Tea and coffee extracts were used to synthesize Pd nanoparticles. Palladium nanoparticles in the range of 20–60 nm were formed (Nadagouda and Varma 2008). *Diopyros kaki* Leaf, *C. Camphora* leaf, *Curcuma longa* tuber and banana peel extracts, were also utilized for the formation of palladium nanoparticles (Sathishkumar et al. 2009; Yang et al. 2010; Song et al. 2010; Bankar et al. 2010).

3.15 Platinum Nanoparticles

Platinum is one of the most costly and novel metal. Its nanoparticles are commonly used as suspensions or colloid in solutions. Platinum nanoparticles are obtaining much attention as they have potent antioxidant activities. These nanoparticles also considered useful for tumor treatment and are currently being used for cancer therapy (Wheate et al. 2010). Platinum nanoparticles have been used in various fields as they possess many unique characteristics. They have great resistance to chemicals and decay so play essential role in the industrial processes (Kang et al. 2008; Wen et al. 2008).

Biosynthesis of platinum nanoparticles was achieved using green tea extract. Two types of nanoparticles were prepared (PtNPs and PtAu-NPs). Both nanoparticles were characterized and used for waste water remediation where the results showed that PtNPs have great potential in water purification. Biogenic fabrication of platinum nanoparticles was studied using aqueous leaf extract of *Cerbera manghas*. Crystalline shaped nanoparticles with small size (9.60–11.70 nm) were observed. FTIR results revealed that alcoholic and carboxyl groups of bioactive compounds were acted as reducing agents. The synthesized particles showed good activity against tested bacterial species.

Cacumen Platycladi plant extract was employed for the biofabrication of platinum nanoparticles. Results revealed that flavonoids and sugars of the plant extract acted as reducing agents. Platinum nanoparticles with extremely small size (0.8–2.4 nm) were synthesized (Zheng et al. 2013).

Leaf extract of *Anacardium occidentale* was used to fabricate platinum nanoparticles where polyols were found to be responsible for the reduction of metal ions and proteins acted as stabilizing agents. Irregular platinum nanoparticles were obtained. *Annona squamosa* was also used for the biosynthesis of Pt nanoparticles. Prepared nanoparticles showed effective larvicidal and insecticidal activity (Kalaiselvi et al. 2015).

Table 3.2 Use of various plants in the biosynthesis of copper, platinum and palladium nanoparticles (Shobha et al. 2014; Kavitha et al. 2013)

Sr.#	Plants	Size (nm)	Nanoparticles
1	*Cinnamomum zeylanicum*	15–20	Pd
2	*Cinnamomum camphora L*	3.2–6	Pd
3	*Gardenia jasminoides*	3–5	Pd
4	*Jatropha curcas L.*	10–12.5	Pb
5	*Soybean (Glycine Max) L.*	15	Pd
6	*Vitus vinifera L.*	661	Pb
7	*Diopyros kaki*	2–12	Pt
8	*Ocimum sanctum*	23	Pt
9	*Calotropis procera L*	1.5–17	Cu
10	*Euphorbia nivulia Latex*		Cu
11	*Gum karaya*	1.6–4.8	Cu
12	*Helianthus annus*	–	Cu
13	*Lantana camara*	20	Cu
14	*Magnolia*	40–100	Cu
15	*Magnolia kobus*	37–110	Cu
16	*Medicago sativa*	–	Cu
17	*Ocimum sanctum*	77	Cu
18	*Soy beans*	–	Cu
19	*Syzygium aromaticum*	40–45	Cu
20	*Tridax procumbens*	–	Cu
21	*Zingiber officinale*	40–25	Cu

Synthesis of platinum nanoparticles was also reported from *Ocimum sancum* leaf extract. Nanoparticles with an average size of 23 nm were formed (Soundarrajan et al. 2012). At first, biosynthesis of platinum nanoparticles was reported using aqueous leaf extract of *Diopyros kaki.* Different concentrations of leaf extract were used. Mean diameter of nanoparticles was calculated as 2–12 nm (Jia et al. 2009). Biofabrication of platinum nanoparticles from plant metabolites was done using *Terminalia chebula.* Polyphenols were found to be responsible for reduction of metal ions. Nanoparticles less than 4 nm were formed (Kumar et al. 2013). Herbal *Bidens tripartitus* extract was tested for its potential to synthesize platinum nanoparticles. A variety of techniques were used to characterize these nanoparticles (Dobrucka 2016) (Table 3.2).

3.16 Selenium Nanoparticles

Selenium is an important micro nutrient, vital compound for numerous enzymes as it activates enzymes. Selenium salts (selenite and selenates) with reducing agents are used to produce Se nanoparticles (Husen and Siddiqi 2014). Biosynthesis of Se nanoparticles can be performed by the reduction of selenium salts in plant extracts (Cox and Bastiaans 2007).

Biogenic fabrication of Se nanoparticles was achieved from *Vitis vinifera* (Sharma et al. 2014). Se nanoparticles (nanorods, nanoballs) were synthesized when aqueous extract of *Capsicum annum* was treated with the H_2SeO_3. Proteins of *C. annum* acted as capping agents in the formation of nanoparticles (Li et al. 2007).

Lemon leaf extract, *Terminalia arjuna* and *Gum arabic* were also exploited for the biofabrication of stable Se nanoparticles (Prasad et al. 2013; Kong et al. 2014). Aqueous seed extract of fenugreek was used to synthesize Se nanoparticles. Synthesized nanopartiles showed significant results when tested against the cancerous cells (Ramamurthy et al. 2013). Spherical shaped nanoparticles with average size 70 nm were synthesized from the peel extract of *citrus reticulate* (Shah et al. 2015).

3.17 Titanium Nanoparticles

Titanium is a compound of enormous importance in different fields of science. It has a great use in pharmaceutical and biomedical sciences due to its safe and biocompatible nature (Gerhardt et al. 2007).

Titanium oxides are nanoparticles of oxides of titanium occur in nature as minerals. These are nonflammable and non ignitable (Newman et al. 2003). Titanium dioxide nanomaterials are non toxic and have important applications in water and air remediation, antimicrobial potential and many other industrial applications (Sarathi and Thilagavathi 2009). In current century, titanium oxides are the most vital nanomaterials which have been widely exploited in the manufacturing of various pharmaceuticals products (Trouiller et al. 2009). Titanium oxide nanoparticles have shown very effective bactericidal activity against pathogenic strains. TiO_2 nanoparticles produce reactive oxygen species upon exposure to UV radiations and these ROS species inhibits the bacterial growth (Allahverdiyev et al. 2011). Initial study on the green synthesis of titanium dioxide nanoparticles using *Catharanthus roseus* and *Annona squamosa* was documented (Roopan et al. 2012).

Titanium oxide nanoparticles have been fabricated from a variety of plants. Leaf extract of *Nyctanthes arbor-tristis* was used to synthesize TiO_2 nanoparticles, globular shaped nanoparticles with 100–150 nm diameters were observed. Titanium nanoparticles with smaller size (36–68 nm) were synthesized from *Eclipta prostrata* leaf extract (Sundrarajan and Gowri 2011). Titanium dioxide nanoparticles were formed from *Catharanthus roseus*. The maximum and minimum size of biosynthesized nanoprticles was calculated as 110 and 25 nm with irregular shape. These green synthesized nanoparticles proved to be effective against *Hippobosca maculate* (Velayutham et al. 2012). *Psidium guajava* was used for the biofabrication of TiO_2 nanoparticles. The antioxidant and antibacterial activity of resulted nanoparticles was tested against *Escherichia coli, Pseudomonas aeruginosa, Staphylococcus aureus* and *Aeromonas hydrophila*. TiO_2 nanoparticles showed significant activity against bacteria (Santhoshkumar et al. 2014).

Numerous plants including lemon grass, tulsi, corriandrum, goose berry have been used for the biosynthesis of titanium oxide nanoparticles (Sundrarajan and Gowri 2011).

TiO_2 nanoparticles have become an advanced group of novel nanomaterials as they possess excellent properties. It is one of the greatly exploited nanomaterials in many fields (Seabra and Durán 2015). Aqueous extract of *Curcuma longa* was tested for the biosynthesis of titanium oxide nanoparticles. Synthesis of nanoparticles was confirmed using XRD. TiO_2 nanoparticles exhibited effective antifungal activity. *Euphorbia prostrate* leaf extract was also investigated for the formation of titanium oxide nanoparticles (Santhoshkumar et al. 2014).

3.18 Zinc Oxide Nanoparticles

Zinc is an essential metal of earth crust. Zinc oxides are inorganic oxides of zinc exists in nature and show a variety of nanomaterials (Szabó et al. 2003). These are approved "generally recognized as safe (GRAS)" by the United States Food and Drug Administration (Premanathan et al. 2011).

The green synthesis of ZnO nanoparticles using plants is gaining great importance because of its numerous useful applications in various fields of science. These are used as bactericidal agents as they have very small size and larger surface area. In fact these metal oxides produce reactive oxygen species that are basically responsible for their improved antibacterial potential (Molina et al. 2006).

Green synthesis of ZnO nanoparticles has been investigated using various plants. Biosynthesis of zinc nanoparticles was reported using *Azadirachta indica*. Phytochemicals of *A. indica* leaf extract acted as reducing and stabilizing agents in nanoparticles formation. Nanoparticles with 50 μm diameter were formed. Fourier transformation infrared spectroscopic results revealed that alcohols, aldehydes, terpenoids and carboxylic acid were the main capping agents in nanoparticles synthesis (Noorjahan et al. 2015).

Zinc oxide nanoparticles were fabricated through the aqueous leaf extract of *Camellia sinesis*. Particle size of synthesized zinc oxide nanoparticles was calculated using particle size analyzer while UV- visible spectrophotometric analysis and SEM was performed for the confirmation of nanoparticlse synthesis. Antimicrobial activity of resultant zinc nanoparticles was also tested (Shah et al. 2015). *Coriandrum sativum* and *Calotropis gigantean* leaves extracts were used for the formation of zinc oxide nanoparticles. Zinc acetate hydrate was added as precursor for the synthesis of zinc oxide nanoparticles (Anastas and Warner 1998).

Recently, the synthesis of ZnO nanoparticles from an important therapeutic plant (*Micrococca mercurialis*) was proposed. Aqueous extracts of fruit, leaf, stem and roots were assessed for the biogenic synthesis of zinc oxide nanoparticles. Synthesized nanoparticles were confirmed by using different techniques. Maximum size (311 nm) of nanoparticles was observed from root extract while stem extract synthesized nanoparticles with 299 nm size (Manokari and Shekhawat 2016). Globular shaped and extremely stable ZnO nanoparticles were produced by using *Aloe vera* extract (Sangeetha et al. 2011).

ZnO nanoparticles have major applications in biomedical science; these are extensively used in cancer therapy (Hanley et al. 2008).

Zinc oxide nanomaterials have ability to increase the growth rate and production of economical food crops. Applications of these nanoparticles to plants enhance the germination, growth and yield of plants (Prasad et al. 2012). Antibacterial potential of zinc nanoparticles makes them excellent nanostructures in the preparation of antibacterial drugs, food preservers, biomedicines, disinfectants and cosmetics (Sharma et al. 2010; Van Dijken et al. 2000).

3.19 Applications of Green Nanoparticles

Nanoparticles have fascinated and prospective exceptional applications in several fields science for instance biotechnology, catalysis, chemical industry, electro- optical devices, electronics and biomedical application (Colvin 2003). Nanoparticles of different sizes and shapes are cytotoxic and exhibit precise responses when come in contact to cells (Huang and El-Sayed 2010).

In many research reports it has been illustrated that silver and gold nanoparticles have antibacterial, antiviral and anti-cancerous potential (Rogers et al. 2008; Gulrajani et al. 2008). Plants have been examined to demonstrate the biological activity of silver NPs. *Citrus limon* was used for the bio-fabrication of silver nanoparticles, the resulted nanoparticles were proved to be effective against *Alternaria brassicicola* and *Fusarium oxysporum*. Antifungal activity of synthesized particles was due to the essential oils of *C. limon* (Vankar and Shukla 2012).

Calotropis gigantean leaf extract was used for the synthesis of silver nanoparticles-particles where the antibacterial activity of the synthesized nanoparticles was tested against *Vibrio alginolyticus*. Zone of inhibition 16 mm for silver nanoparticles and 20 mm was measured for silver nitrate solution while simple leaf extract was ineffective against tested bacteria. Biosynthesized silver nanoparticles can also act as valuable substitute in fighting against resistant pathogenic bacteria. Green synthesized silver NPs using *Sargassum tenerrimum* exhibited remarkable activity due to the presence of plant phyto-constituents against gram positive and gram negative bacteria (Baskaralingam et al. 2012; Kumar et al. 2012).

Plant synthesized silver NPs have been reported for antifungal activity, activity against parasites and HIV- virus (Marimuthu et al. 2011) (Table 3.3).

Table 3.3 Major applications of metal nanoparticles

Sr.#	Metal nanoparticles	Applications	References
1	Silver	Antimicrobial, anticancerous	Huang et al. (2007)
2	Gold	Anticancerous, antimicrobial	Cai and Chen (2007)
3	Palladium	Antibacterial	Amarnath et al. (2012)
4	Selenium	Anticancerous	Prasad et al. (2012)
5	Copper	Antimicrobial, anticancerous	Lee et al. (2011)
6	Zinc oxide	Cosmetics, coating	Mittal et al. (2013)
7	Iron oxide	Anticancerous	Mittal et al. (2013)
8	Platinum	Anticancerous	Mittal et al. (2013)

3.20 Conclusion

In the field of Nano-biotechnology, green synthesis of metal nanoparticles has increased for the fabrication of new nano scale materials that are environment friendly, less expensive, stable nanoparticles with an immense significance for many applications in different areas of science including medicine, pharmaceutics, cancer control and agriculture. In the current century, nanotechnology stimulating the advancement in all areas of life, hence biological methods for nanoparticles fabrication will arise as ecofriendly and safer choice than the usual methods. The different biological materials have been used for the synthesis of nanoparticles; exploitation of plants for the easy and vigorous production of nanoparticles is a marvelous approach. With the vast plant variety much more plant species are in way to be exploited and reported in future towards fast and single step procedure with green principle.

References

Allahverdiyev AM et al (2011) Antimicrobial effects of TiO_2 and Ag_2O nanoparticles against drug-resistant bacteria and leishmania parasites. Future Microbiol 6(8):933–940

Amarnath K et al (2012) Synthesis and characterization of chitosan and grape polyphenols stabilized palladium nanoparticles and their antibacterial activity. Coll Surf B Biointerfaces 92:254–261

Anastas PT, Warner JC (1998) Principles of green chemistry. Green chemistry. Theory Pract:29–56

Ashoori R (1996) Electrons in artificial atoms. Nature 379:413–419

Awwad AM, Salem NM (2012) Green synthesis of silver nanoparticles by mulberry leaves extract. Nanosci Nanotechnol 2(4):125–128

Balamurugan M, Saravanan S, Soga T (2014) Synthesis of iron oxide nanoparticles by using *Eucalyptus globulus* plant extract. E-J Surf Sci Nanotech 12:363–367

Bankar A et al (2010) Banana peel extract mediated novel route for the synthesis of silver nanoparticles. Coll Surf A Physicochem Eng Aspects 368(1):58–63

Baskaralingam V et al (2012) Green synthesis of silver nanoparticles through *Calotropis gigantea* leaf extracts and evaluation of antibacterial activity against *Vibrio alginolyticus*. Nanotech Develop 2(1):3

Bencherif SA et al (2009) Nanostructured hybrid hydrogels prepared by a combination of atom transfer radical polymerization and free radical polymerization. Biomaterials 30(29):5270–5278

Berry CC, Curtis AS (2003) Functionalisation of magnetic nanoparticles for applications in biomedicine. J Phy D Appl Phy 36(13):198

Bhau B et al (2015) Green synthesis of gold nanoparticles from the leaf extract of *Nepenthes khasiana* and antimicrobial assay. Adv Mater Lett 6:55–58

Bhosale R, et al. (2014) Innovative eco-friendly approaches for green synthesis of silver nanoparticles. Int J Pharm Sci Nanotech7:2328–2337

Boudad H et al (2001) Combined hydroxypropyl-β-cyclodextrin and poly (alkylcyanoacrylate) nanoparticles intended for oral administration of saquinavir. Int J Pharm 218(1):113–124

Buhleier E, Wehner W, Vögtle F (1978) Cascade and nonskid chain like syntheses of molecular cavity topologies. Chemischer Informationsdienst 9(25)

Cai W, Chen X (2007) Nanoplatforms for targeted molecular imaging in living subjects. Small 3(11):1840–1854

Chah S, Hammond MR, Zare RN (2005) Gold nanoparticles as a colorimetric sensor for protein conformational changes. Chem Biol 12(3):323–328

Chandran K, Song S, Yun SI (2014) Effect of size and shape controlled biogenic synthesis of gold nanoparticles and their mode of interactions against food borne bacterial pathogens. Arab J Chem https://doi.org/10.1016/j.arabjc.2014.11.041

Cheng L et al (2012) Multifunctional nanoparticles for upconversion luminescence/MR multimodal imaging and magnetically targeted photothermal therapy. Biomaterials 33(7):2215–2222

Chu C et al (1988) J Trauma Inj Infect. Crit Care 28:1488–1492

Coll JL (2011) Cancer optical imaging using fluorescent nanoparticles. Nanomed 6(1):7–10

Colvin VL (2003) The potential environmental impact of engineered nanomaterials. Nat Biotech 21(10):1166–1170

Connor EE et al (2005) Gold nanoparticles are taken up by human cells but do not cause acute cytotoxicity. Small 1(3):325–327

Cornell RM, Schwertmann U (2003) The iron oxides: structure, properties, reactions, occurrences and uses. Wiley, Weinheim

Cox DN, Bastiaans K (2007) Understanding Australian consumers' perceptions of selenium and motivations to consume selenium enriched foods. Food Quality Pref 18(1):66–76

Daisy P, Saipriya K (2012) Biochemical analysis of Cassia fistula aqueous extract and phytochemically synthesized gold nanoparticles as hypoglycemic treatment for diabetes mellitus. Int J Nanome D7:1189–1202

Daniel SK et al (2013) Biosynthesis of Cu, ZVI, and Ag nanoparticles using *Dodonaea viscosa* extract for antibacterial activity against human pathogens. J Nanopar Res 15(1):1–10

Dhuper S, Panda D, Nayak P (2012) Green synthesis and characterization of zero valent iron nanoparticles from the leaf extract of *Mangifera indica*. Nano Trends J Nanotech App 13(2):16–22

Dobrucka R (2016) Synthesis and structural characteristic of platinum nanoparticles using herbal *Bidens Tripartitus* extract. J Inorg Organometall Polymers Mat 26(1):219–225

Edison TJI, Sethuraman M (2012) Instant green synthesis of silver nanoparticles using *Terminalia chebula* fruit extract and evaluation of their catalytic activity on reduction of methylene blue. Process Biochem 47(9):1351–1357

El-Sayed IH, Huang X, El-Sayed MA (2006) Selective laser photo-thermal therapy of epithelial carcinoma using anti EGFR antibody conjugated gold nanoparticles. Cancer Lett 239(1):129–135

Elumalai E et al (2010) Green synthesis of silver nanoparticle using *Euphorbia hirta* L and their antifungal activities. Arch Appl Sci Res 2(6):76–81

Ganeshkumar M et al (2012) Sun light mediated synthesis of gold nanoparticles as carrier for 6-mercaptopurine: preparation, characterization and toxicity studies in zebrafish embryo model. Mat Res Bull 47(9):2113–2119

Ganeshkumar M et al (2013) Spontaneous ultra fast synthesis of gold nanoparticles using *Punica granatum* for cancer targeted drug delivery. Coll Surf B Biointerfaces 106:208–216

Gardea-Torresdey JL et al (2003) Alfalfa sprouts: a natural source for the synthesis of silver nanoparticles. Langmuir 19(4):1357–1361
Geetha N et al (2012) Biofabrication of silver nanoparticles using leaf extract of *Chromolaena Odorata* (L.) King and Robinson. An international conference on nuclear energy. Environm Biol Sci 8:56–59
Gerhardt LC, Jell G, Boccaccini A (2007) Titanium dioxide (TiO2) nanoparticles filled poly (D, L lactid acid)(PDLLA) matrix composites for bone tissue engineering. J Mat Sci Mat Med 18(7):1287–1298
Gong P et al (2007) Preparation and antibacterial activity of Fe_3O_4@ Ag nanoparticles. Nanotech 18(28):285604
Gonzalez A, Noguez C (2007) Influence of morphology on the optical properties of metal nanoparticles. J Comp Theoretical Nanosci 4(2):231–238
Gulrajani M et al (2008) Preparation and application of silver nanoparticles on silk for imparting antimicrobial properties. J Appl Polym Sci 108(1):614–623
Hanley C et al (2008) Preferential killing of cancer cells and activated human T cells using ZnO nanoparticles. Nanotech 19(29):295103
Hayashi H et al (2004) pH-sensitive nanogel possessing reactive PEG tethered chains on the surface. Macromolecules 37(14):5389–5396
Honary S et al (2012) Green synthesis of copper oxide nanoparticles using *Penicillium aurantiogriseum, Penicillium citrinum and Penicillium waksmanii*. Dig J Nanomater Bios 7:999–1005
Huang X, El-Sayed MA (2010) Gold nanoparticles: optical properties and implementations in cancer diagnosis and photothermal therapy. J Adv Res 1(1):13–28
Huang J et al (2007) Biosynthesis of silver and gold nanoparticles by novel sundried *Cinnamomum camphora* leaf. Nanotechology 18(10):105104
Huber DL (2005) Synthesis, properties, and applications of iron nanoparticles. Small 1(5):482–501
Husen A, Siddiqi KS (2014) Plants and microbes assisted selenium nanoparticles: characterization and application. J Nanobiotech 12(1):1
Janka J, Maldarelli F (2004) Prion diseases: update on mad cow disease, variant creutzfeldt-jakob disease, and the transmissible spongiform encephalopathies. Curr Inf Dis Rep 6(4):305–315
Jayandran M, Haneefa MM, Balasubramanian V (2015) Green synthesis and characterization of manganese nanoparticles using natural plant extracts and its evaluation of antimicrobial activity. J Appl Pharm Sci 5(12):105–110
Jayaseelan C et al (2013) Green synthesis of gold nanoparticles using seed aqueous extract of *Abelmoschus esculentus* and its antifungal activity. Ind Crop Prod 45:423–429
Jeong SH, Yeo SY, Yi SC (2005) The effect of filler particle size on the antibacterial properties of compounded polymer/silver fibers. J Mat Sci 40(20):5407–5411
Jia L et al (2009) The biosynthesis of palladium nanoparticles by antioxidants in *Gardenia jasminoides* Ellis: long lifetime nanocatalysts for p-nitrotoluene hydrogenation. Nanotech 20(38):385601
Kabanov AV, Vinogradov SV (2009) Nanogels as pharmaceutical carriers: finite networks of infinite capabilities. Angewandte Chemie Int Ed 48(30):5418–5429
Kalaiselvi A et al (2015) Synthesis and characterization of palladium nanoparticles using *Catharanthus roseus* leaf extract and its application in the photo-catalytic degradation. Spectro Chimica Acta Part A Mol Biomol Spec 135:116–119
Kale A et al (2013) Directed self-assembly of CdS quantum dots on bacteriophage P22 coat protein templates. Nanotech 24(4):045603
Kaler A et al (2011) Extracellular biosynthesis of silver nanoparticles using aqueous extract of *Candida viswanathii*. J Bionanosci 5(1):53–58
Kang X et al (2008) Glucose biosensors based on platinum nanoparticles-deposited carbon nanotubes in sol–gel chitosan/silica hybrid. Talanta 74(4):879–886
Kavitha K et al (2013) Plants as green source towards synthesis of nanoparticles. Int Res J Biol Sci 2(6):66–76

Kaviya S et al (2011) Biosynthesis of silver nanoparticles using Citrus sinensis peel extract and its antibacterial activity. Spectrochimica Acta Part A Mol Biomol Spect 79(3):594–598
Kievit FM, Zhang M (2011) Surface engineering of iron oxide nanoparticles for targeted cancer therapy. Acc Chem Res 44(10):853–862
Kl N et al (2014) Green synthesis and characterization of silver nanoparticles from aqueous extract of *Basella alba* and their in-vitro antioxidant potentials. Int J Pharm Pharmaceut Sci 6(10):393–396
Kong H et al (2014) Synthesis and antioxidant properties of gum arabic-stabilized selenium nanoparticles. Int J Biol Macromol 65:155–162
Koo B et al (2012) Hollow iron oxide nanoparticles for application in lithium ion batteries. Nano Lett 12(5):2429–2435
Krishnaraj C et al (2014) Acalypha indica Linn: biogenic synthesis of silver and gold nanoparticles and their cytotoxic effects against MDA-MB-231, human breast cancer cells. Biotech Rep 4:42–49
Kulkarni VD, Kulkarni PS (2013) Green synthesis of copper nanoparticles using *Ocimum sanctum* leaf extract. Int J Chem Stud 1(3):1–4
Kulkarni N, Muddapur U (2014) Biosynthesis of metal nanoparticles: a review. J of Nanotech 2014:1–8
Kumar V, Yadav SK (2009) Plant mediated synthesis of silver and gold nanoparticles and their applications. J Chem Tech Biotech 84(2):151–157
Kumar A et al (2008) Silver-nanoparticle-embedded antimicrobial paints based on vegetable oil. Nat Mat 7(3):236–241
Kumar VG et al (2011) Facile green synthesis of gold nanoparticles using leaf extract of antidiabetic potent *Cassia auriculata*. Coll Surf B: Biointerfaces 87(1):159–163
Kumar P et al (2012) Synthesis of silver nanoparticles from *Sargassum tenerrimum* and screening phytochemicals for its antibacterial activity. Nano Biomed Eng 4(1):12–16
Kumar KM et al (2013) Biobased green method to synthesise palladium and iron nanoparticles using *Terminalia chebula* aqueous extract. Spectrochimica Acta Part A Mol Biomol Spectrosc 102:128–133
Latha N, Gowri M (2014) Bio synthesis and characterisation of Fe_3O_4 nanoparticles using *Caricaya Papaya* leaves extract. Synthesis 12:17
Lee H, Lee G, Na RJ, Jung HY, Jae YS, Beom SK (2011) Biological synthesis of copper nanoparticles using plant extract. NSTI Nanotech 3(1):371–374
Li YP et al (2001) PEGylated polycyanoacrylate nanoparticles as tumor necrosis factor-α carriers. J Control Release 71(3):287–296
Li L et al (2006) Synthesis, properties, and environmental applications of nanoscale iron-based materials: a review. Crit Rev Env Sci Tech 36(5):405–431
Li S et al (2007) Rapid, room temperature synthesis of amorphous selenium/protein composites using *Capsicum annuum* L extract. Nanotech 18(40):405101
Lynch N et al (2006) PANDAS (Paediatric autoimmune neuropsychiatric disorder associated with streptococcal infection). Irish Med J 99(5):155–155
Majumder D (2012) Bioremediation: copper nanoparticles from electronic-waste. Intl J Eng Sci Tech 4(10)
Manokari M, Shekhawat MS (2016) Green synthesis of zinc oxide nanoparticles using plant extracts of *Leucas aspera* (Willd.) L. Int J Biol Pap 1(1):22–27
Marimuthu S et al (2011) Evaluation of green synthesized silver nanoparticles against parasites. Parasit Res 108(6):1541–1549
Marshall AT et al (2007) Accumulation of gold nanoparticles in *Brassica juncea*. Int J Phytoremediation 9(3):197–206
McKenzie LC, Hutchison JE (2004) Green nanoscience. Chim Oggi 22(9):30–33
Melchert WR, Reis BF, Rocha FR (2012) Green chemistry and the evolution of flow analysis. A review. Ana Chim Acta 714:8–19

Meyers MA, Mishra A, Benson DJ (2006) Mechanical properties of nanocrystalline materials. Prog Mater Sci 51(4):427–556

Mishra V, Sharma R, Nakuleshawar D, Gupta K (2014) A review on green synthesis of nanoparticles and evaluation of antimicrobial activity. Inter J Green Herb Chem 3(1):81–94

Mittal AK, Chisti Y, Banerjee UC (2013) Synthesis of metallic nanoparticles using plant extracts. Biotech Adv 31(2):346–356

Mody VV et al (2010) Introduction to metallic nanoparticles. J Pharm Bioallied Sci 2(4):282

Molina MA, Ramos JL, Espinosa M (2006) Urgel, a two partner secretion system is involved in seed and root colonization and iron uptake by *Pseudomonas putida* KT2440. Env Microbiol 8(4):639–647

Mukherjee P et al (2001) Fungus-mediated synthesis of silver nanoparticles and their immobilization in the mycelial matrix: a novel biological approach to nanoparticle synthesis. Nano Lett 1(10):515–519

Mukunthan K, Balaji S (2012) Cashew apple juice (Anacardium occidentale L.) speeds up the synthesis of silver nanoparticles. Int J Green Nanotech 4(2):71–79

Murray CB, Kagan C, Bawendi M (2000) Synthesis and characterization of monodisperse nanocrystals and close-packed nanocrystal assemblies. Ann Rev Mat Sci 30(1):545–610

Murugan K, et al. (2015) *Cymbopogon citratus*-synthesized gold nanoparticles boost the predation efficiency of copepod Mesocyclops aspericornis against malaria and dengue mosquitoes. Exp Parasitol153:129–138

Nadagouda MN, Varma RS (2008) Green synthesis of silver and palladium nanoparticles at room temperature using coffee and tea extract. Green Chem 10(8):859–862

Naika HR et al (2015) Green synthesis of CuO nanoparticles using *Gloriosa superba* L. extract and their antibacterial activity. J Taibah Uni Sci 9(1):7–12

Nanjwade BK et al (2009) Dendrimers: emerging polymers for drug-delivery systems. Eur J Pharm Sci 38(3):185–196

Naseem T, Farrukh MA (2015) Antibacterial activity of green synthesis of iron nanoparticles using *Lawsonia inermis and Gardenia jasminoides* leaves extract. J Chem 2015:1–7

Nasrollahzadeh M, Sajadi SM (2015) Green synthesis of copper nanoparticles using *Ginkgo biloba* L. leaf extract and their catalytic activity for the Huisgen [3+ 2] cycloaddition of azides and alkynes at room temperature. J Coll Interface Sci 457:141–147

Nath D, Banerjee P (2013) Green nanotechnology–a new hope for medical biology. Environ Toxicol Pharmacol 36(3):997–1014

Noorjahan C et al (2015) Green synthesis and characterization of zinc oxide nanoparticles from neem (*Azadirachta indicia*). Int J Sci Eng Tech Res 4(30):5751–5753

Pal SL, Jana U, Manna PK, Mohanta GP, Manavalan R (2011) Nanoparticle: an overview of preparation and characterization. J Appl Pharm Sci 01(06):228–234

Pandey S et al (2012) Green synthesis of highly stable gold nanoparticles using *Momordica charantia* as nano fabricator. Arch Appl Sci Res 4(2):1135–1141

Parak WJ et al (2003) Biological applications of colloidal nanocrystals. Nanotechology 14(7):R15

Park Y et al (2011) Polysaccharides and phytochemicals: a natural reservoir for the green synthesis of gold and silver nanoparticles. IET Nanobiotech 5(3):69–78

Passam HC et al (2007) A review of recent research on tomato nutrition, breeding and post-harvest technology with reference to fruit quality. Eur J Plant Sci Biotech 1(1):1–21

Pattanayak M, Nayak P (2013) Ecofriendly green synthesis of iron nanoparticles from various plants and spices extract. Int J Plant Ani Env Sci 3(1):68–78

Philip D (2010) Green synthesis of gold and silver nanoparticles using *Hibiscus rosasinensis*. Phys E Low-Dimen Sys Nanostr 42(5):1417–1424

Pillai RK et al (2012) Vermifugal activity of biofabricated silver nanoparticles. Res J Recent Sci 1:47–51

Prabhu S, Poulose EK (2012) Silver nanoparticles: mechanism of antimicrobial action, synthesis, medical applications, and toxicity effects. Int Nano Lett 2(1):1–10

Prasad T et al (2012) Effect of nanoscale zinc oxide particles on the germination, growth and yield of peanut. J Plant Nut 35(6):905–927
Prasad KS et al (2013) Biosynthesis of se nanoparticles and its effect on UV-induced DNA damage. Coll Surf B: Biointerfaces 103:261–266
Premanathan M et al (2011) Selective toxicity of ZnO nanoparticles toward Gram-positive bacteria and cancer cells by apoptosis through lipid peroxidation. Nanomed Nanotech Biol Med 7(2):184–192
Ramamurthy Ch, Sampath KS P Arunkumar, M Suresh Kumar, V Sujatha, K Premkumar, C Thirunavukkarasu (2013) Green synthesis and characterization of selenium nanoparticles and its augmented cytotoxicity with doxorubicin on cancer cells. Bioprocess Biosyst Eng 36(8):1131–1139
Raveendran P, Fu J, Wallen SL (2003) Completely "green" synthesis and stabilization of metal nanoparticles. J Amr Chem Soc 125(46):13940–13941
Ravishankar RV, Jamuna B (2011) Nanoparticles and their potential application as antimicrobials. Science against microbial pathogens, communicating current research and technological advances. Formatex Badajoz, pp 197–209
Rogers JV et al (2008) A preliminary assessment of silver nanoparticle inhibition of monkeypox virus plaque formation. Nanoscale Res Lett 3(4):129
Roopan SM et al (2012) Efficient phyto-synthesis and structural characterization of rutile TiO_2 nanoparticles using *Annona squamosa* peel extract. Spectrochimica Acta Part A Mol Biomol Spect 98:86–90
Roya K, Sarkar C, Ghosh C (2015) Single-step novel biosynthesis of silver nanoparticles using Cucumis sativus fruit extract and study of its photcatalytic and antibacterial activity. Digest J Nanomat Biostr 10(1):107
Safaepour M et al (2009) Green synthesis of small silver nanoparticles using geraniol and its cytotoxicity against fibrosarcoma-wehi 164. Avicenna J Med Biotech 1(2):111–115
Sangeetha G, Rajeshwari S, Venckatesh R (2011) Green synthesis of zinc oxide nanoparticles by *Aloe barbadensis* Miller leaf extract: structure and optical properties. Mat Res Bull 46(12):2560–2566
Santhoshkumar T et al (2014) Green synthesis of titanium dioxide nanoparticles using *Psidium guajava* extract and its antibacterial and antioxidant properties. Asi Pac J Ttrop Med 7(12):968–976
Sarathi P, Thilagavathi G (2009) Synthesis and characterization of titanium dioxide nano-particles and their applications to textiles for microbe resistance. J Textile Apparel Tech Manag 6(2)
Sathishkumar M et al (2009) Phyto-crystallization of palladium through reduction process using *Cinnamom zeylanicum* bark extract. J Hazard Mat 171(1):400–404
Seabra AB, Durán N (2015) Nanotoxicology of metal oxide nanoparticles. Metals 5(2):934–975
Senthil M, Ramesh C (2017) Biogenic synthesis of Fe_3O_4 nanoparticles using Tridax Procumbens leaf extract and its antibacterial activity on *Pseudomonas aeruginosa*. Digest J Nanomat Biostr 7(4)
Sepeur S (2008) Nanotechnology: technical basics and applications. Vincentz Network GmbH & Co KG, Hannover
Shah S et al (2014) Green synthesis of iron nanoparticles using plant extracts. Int J Biol Pharm Res 5:549–552
Shah M et al (2015) Green synthesis of metallic nanoparticles via biological entities. Mat 8(11):7278–7308
Shankar SS et al (2003) Bioreduction of chloroaurate ions by Geranium leaves and its endophytic fungus yields gold nanoparticles of different shapes. J Mat Chem 13(7):1822–1826
Sharma VK, Yngard RA, Lin Y (2009) Silver nanoparticles: green synthesis and their antimicrobial activities. Adv Colloid Interface Sci 145(1):83–96
Sharma D et al (2010) Synthesis of ZnO nanoparticles and study of their antibacterial and antifungal properties. Thin Solid Films 519(3):1224–1229

Sharma G et al (2014) Biomolecule-mediated synthesis of selenium nanoparticles using dried *Vitis vinifera* (raisin) extract. Mol 19(3):2761–2770
Shobha G, Moses V, Ananda S (2014) Biological synthesis of copper nanoparticles and its impact. Int J Pharm Sci Invent 3(8):6–28
Si S, Mandal TK (2011) Tryptophan based peptides to synthesize gold and silver nanoparticles: a mechanistic and kinetic study. Chem A Eur J 13(11):3160–3168
Singh M et al (2008) Nanotechnology in medicine and antibacterial effect of silver nanoparticles. Digest J Nanomat Biostr 3(3):115–122
Singh C et al (2011) A green biogenic approach for synthesis of gold and silver nanoparticles using *Zingiber officinale*. Dig J Nanomat Biostr 6(2):535–542
Singh A, Singh S, Singh N (2014) Scientia research library. J Appl Chem 2(2):77–82
Slomkowski S et al (2011) Terminology of polymers and polymerization processes in dispersed systems (IUPAC recommendations 2011). Pure Appl Chem 83(12):2229–2259
Song JY, Jang HK, Kim BS (2009) Biological synthesis of gold nanoparticles using *Magnolia kobus and Diopyros kaki* leaf extracts. Process Biochem 44(10):1133–1138
Song JY, Kwon EY, Kim BS (2010) Biological synthesis of platinum nanoparticles using *Diopyros kaki* leaf extract. Bioprocess Biosyst Eng 33(1):159–164
Soundarrajan C et al (2012) Rapid biological synthesis of platinum nanoparticles using *Ocimum sanctum* for water electrolysis applications. Bioprocess Biosyst Eng 35(5):827–833
Srinivasan M, Rajabi M, Mousa SA (2015) Multifunctional nanomaterials and their applications in drug delivery and cancer therapy. Nanomaterials 5(4):1690–1703
Subhankari I, Nayak P (2013) Synthesis of copper nanoparticles using *Syzygium aromaticum* (Cloves) aqueous extract by using green chemistry. World J Nano Sci Tech 2(1):14–17
Sumit S, Nayak P (2012) Green synthesis of gold nanoparticles using various extract of plants and spices. Int J Sci Innov Discov 2(3):325–350
Sundrarajan M, Gowri S (2011) Green synthesis of titanium dioxide nanoparticles by *Nyctanthes arbor-tristis* leaves extract. Chalcogenide Lett 8(8):447–451
Szabó T, Németh J, Dékány I (2003) Zinc oxide nanoparticles incorporated in ultrathin layer silicate films and their photocatalytic properties. Coll Surf A Physicochem Eng Asp 230(1):23–35
Thakkar KN, Mhatre SS, Parikh RY (2010) Biological synthesis of metallic nanoparticles. Nanomed Nanotech Biol Med 6(2):257–262
Thakor A et al (2011) Gold nanoparticles: a revival in precious metal administration to patients. Nano Lett 11(10):4029–4036
Tian J et al (2007) Topical delivery of silver nanoparticles promotes wound healing. Chem Med Chem 2(1):129–136
Tomar A, Garg G (2013) Short review on application of gold nanoparticles. Global J Pharm 7(1):34–38
Tran QH, Le AT (2013) Silver nanoparticles: synthesis, properties, toxicology, applications and perspectives. Adv Nat Sci Nanosci Nanotech 4(3):033001
Trouiller B et al (2009) Titanium dioxide nanoparticles induce DNA damage and genetic instability in vivo in mice. Cancer Res 69(22):8784–8789
Vadlapudi V et al (2014) Synthesis of green metallic nanoparticles (NPs) and applications. Oriental J Chem 29(4):1589–1595
Valodkar M et al (2011) Biocompatible synthesis of peptide capped copper nanoparticles and their biological effect on tumor cells. Mat Chem Phys 128(1):83–89
Van Dijken A et al (2000) Identification of the transition responsible for the visible emission in ZnO using quantum size effects. J Luminescence 90(3):123–128
Vankar PS, Shukla D (2012) Biosynthesis of silver nanoparticles using lemon leaves extract and its application for antimicrobial finish on fabric. Appl Nanosci 2(2):163–168
Velayutham K et al (2012) Evaluation of *Catharanthus roseus* leaf extract-mediated biosynthesis of titanium dioxide nanoparticles against *Hippobosca maculata and Bovicola ovis*. Parasitol Res 111(6):2329–2337

Wen Z, Liu J, Li J (2008) Core/shell pt/c nanoparticles embedded in mesoporous carbon as a methanol tolerant cathode catalyst in direct methanol fuel cells. Adv Mat 20(4):743–747

Wheate NJ et al (2010) The status of platinum anticancer drugs in the clinic and in clinical trials. Dalton Trans 39(35):8113–8127

Yang X et al (2010) Green synthesis of palladium nanoparticles using broth of *Cinnamomum camphora* leaf. J Nanopart Res 12(5):1589–1598

Zhang Q, Shen Z, Nagai T (2001) Prolonged hypoglycemic effect of insulin-loaded polybutylcyanoacrylate nanoparticles after pulmonary administration to normal rats. Int J Pharm 218(1):75–80

Zheng B et al (2013) Plant-mediated synthesis of platinum nanoparticles and its bioreductive mechanism. J Coll Interface Sci 396:138–145

Chapter 4
Nanofibers and Nanowires

Aneeqa Sabah and Gabour Loius Hornyak

4.1 Fundamentals of Nanomaterials

Nanomaterials can be categorized into three main types on the basis of dimensions that cannot be confined to nano scale (<100 nm). zero-dimensional (0-D) quantum dots, one- dimensional (1-D) nano wires, nanofibers and tubes, two- dimensional (2-D) super lattices and thin films. This classification of having one or more small and confined dimensions small of nanostructured materials in comparison of bulk, exhibit special properties. In 0-D materials, all three dimensions are restricted to nanometer scale. For example, the nanoparticles in any form, can be amorphous, crystalline, single or poly crystalline, exist in various shapes and sizes. 2-D nanomaterials have two unconfined dimensions and only one is restricted to nanometer scale. 2- D materials include nano plating, nano disks, nano coating, nanolayers and films. They can also be either metallic or crystalline. 2-D materials are formed by depositing some specified thickness of specimen on substrate. Finally, the 1-D materials are those having one dimension out of nanoscale and other two are confined to nano scale with negligible crystallite size. For example, the nano rods, nano wires and fibers and nanotubes are 1-D nanomaterials. Electrons and holes exhibited quantum confinement within two dimensions and free to travel in third one.

A. Sabah (✉)
Department of Physics, Lahore College for Women University, Lahore, Pakistan

G. L. Hornyak
Asian Institute of Technology, Khlong Nueng, Thailand

S. Javad, A. Butt (eds.), *Nanobotany*, https://doi.org/10.1007/978-3-319-77119-9_4

4.2 Introduction to Nano Fibers and Nanowires

The term fiber comes from Latin fibra: meaning: an elongated thread like structure. It is essential to understand first, the viewpoint of various researchers and scientist about fibers. Botanist consider the fibers as elongated, thick walled cells that give support and strength to plant tissues. Fibers are defined as natural and synthetic filaments in textile industry such as cotton and nylon to made different functional materials. Biochemist and physiologist identify the term fiber as indigestible plant matter having cellulose to stimulate intestinal peristalsis.

The term "nano" comes from Greek word,"nannas" meaning "dwarf" small, little entity. Technically the nano defines as a physical quantity of one billionth of meter (10^{-9} m). In nano-bio engineering and research, naturally existing nanofibers are frequently in use to reconstruct the functional nanostructures and devices. Nano and micro wires provide recent breakthrough in fabricating the flexible electronic device by aligning the organic wires on substrate under controlled conditions like flexible displays in computers, transistors, smart phones etc [Queens University Report].

Nanowires are the structures with the length to width ratio (aspect ratio) being greater than 1000. The diameter of wires is restricted to tens of nanometer scale but unconstrained length. Nanofibers are highly preferable materials for variety of applications ranging from textile industry to air-craft structures due to high surface area, flexibility and specific directional strength.

4.3 One-Dimensional Nanomaterials

Nanofibers geometrically (according to shape, size and dimensions) falls into the category of 1-D nanomaterials. Nano rods, nanotubes, nano wires all comes within nanofibers. Rather, nanofibers are much more flexible as can be aligned with highly flexible nano elements like globular proteins e.g. DNA helix strands. One of the best example of nanowires and nanorods is carbon nanotubes also termed as quantum wire (as quantum effect become prominent at nano scale, electron behave like wave.) having two dimensions restricted to exciton Bohr radius (electron and hole pair distance within the conduction and valence band respectively).

4.4 Types of Nanowires

Different types of nanowires and fibers exist including metallic, semiconducting, insulating, molecular either organic or inorganic, bio-templated within the range of few 100 of nm in diameter and exceeding the length over 100 μm. Basically the diameter depicts the dimensions and homogeneous nucleation establish the suitable wire length (Gabor et al. 2008).

Biomimetic Nanofibers/wires: Naturally occurring biological structures exhibit attractive properties that can be implanted within metallic nanoparticles to design biomimetic complex structures. Fungal species occurs in variety of geometrical shapes, patterns, textures as well as colors that is very motivating for engineers, scientists and researchers to be used as templates. The hard surfaces of microorganisms are found to be the best template for self-organization of nanoparticles into uniform nanofiber arrays with advanced functional properties. These bio-nano fibers composites may lead to cost-effective, small size and reliable architecture in comparison to complex and trendy techniques in market. Tip Fungal hyphae tip growth is sensitive to the variety of physical, chemical and environmental factor (Gow 1994). The influence of external stimuli such as light, heat, radiations, nutrients, flow sensing, external fields and gravity to control the direction of growing tip can produce the directed mesh of wires for potential application, specifically in electric circuits. Dimensions of nano and micro wires can be well controlled by tropism of microorganisms towards external stimuli for their survival and reproduction. Cellular tip extension and orientation is controlled by vesicles released by micro-tubules and actin (Alexandra and Gow 2009).

Helical Nanofibers, Dumbbell-shaped molecules self-assembled into helical cylinders of ~ diameter of 8 nm and pitch length of 5.6 nm in an aqueous solution. These elementary helical fibers further assembled into super-helical fibers having length of several micrometers. Such 1-D biomimetic nanofibers may introduce new era in flexible electronic and photonic fields (Bae et al. 2005).

Branched and Bundled Nanofiber, Bundled and branched nanofibers were fabricated by self-assembly of conjugated amphiphilic block co-polymers of polythiophene (PHT) and polyethylene glycol (PEG). The shape, length and density of bundled and branched nanofibers were controlled by the length, ratio block co-polymer and homopolymer and quality of solvent. Self-assembled amphilic polymers can be transformed into isolated, bundled, and branched nanofibers (Kamps et al. 2012). Branched and porous carbon nanofibers are frequently used in electrochemical double-layer capacitors (Tao et al. 2006), ***Hollow and composite*** (McCann et al. 2005; Li and Xia 2004).

Ribbon like nanofibers were also reported (Chang et al. 2011).

4.5 Properties of Nanofibers and Nanowires

Nanowires and nanofibers exhibit many different and interesting properties in comparison of bulk due to quantum confinement.

4.5.1 Mechanical Properties

Nanowires are frequently in use as connecting wires in electronic circuits and devices as well. The use of carbon nanotube as probe tip in AFM or STM is termed as nanostructure but if used in some electronic circuit as connecting conducting wire than referred to working as device. Young's modulus of gold nanowires as nanomechanically measured by AFM tip is found to be independent of diameter and yield strength and provide 100 times better strength than bulk one (Wu et al. 2005). Self-twisting and flexibility of nanofibers is quite impressive for advancement in flexible electronics. Greater tensile strength, flexibility and bending of polymer nano-fibers composite is improved to extensive level by surface functionalizing). Nanofiber may exist as single nanofiber, nanofiber yarn, nanofiber membrane. Fiber diameter, porosity, and pore size, surface contact angle measurementscorrespond to mechanical strength of nano-fibers (Cassie and Baxter 1994).

4.5.2 Optical Properties

Optical properties of nanowires are anisotropic, as orientation play the role. Nanowires can be aligned in the presence of electromagnetic field accordingly. Confined population of electrons is affected by the restoring force the depolarization field. This depolarization effect is caused by surface plasmon of curved surface. The degree of effect is dependent on aspect ratio of synthesized nanoparticles (Pérez-Juste et al. 2005). Longitudinal and transvers, two plasmon modes are observed in absorption spectrum for gold nanorods (Hornyak 1997). Addition of organic materials to nanowires alters their optical response as absorption maxima in UV-visible spectrum shifted to lower energies, the red shifts (Gabor et al. 2008). If the additive one is an insulating material, then polarization is diffuse toward red shift.

4.5.3 Electrical Properties

In nanowires, electrons are free to move without any collisions with impurities, hence no energy loss occur. This effect will become prominent, when the wire width is below the mean free path of electron. For example, the ballistic conduction (electrons are free to move from one electrode to another) or diffusive, depending on elastic scattering and electron tunnelingin single walled carbon nanotubes SWCNT as reported. The nanotubes can withstand with high value of current ~ 1 mA (Chappel and Zabau 2002). Unlike the ballistic transport in CNT, nanowires also exhibit the conductivity strongly dependent on edge effect. Atoms lies on the nanowires surface are loosely bound with neighboring atoms. This dangling bonding creates the defects that may cause the nanowires to conduct electricity. For small size nanowires and nanofibers, the edge effect becomes prominent. Bacterial nanowires with electrical conductivity comparable to semiconductor wires and elasticity are similar to typical polymeric material. These protein-nature bacterial nanowires may provide potential application for bio-fuel cell and bio-nano flexible electronic area.

4.6 Nano Fibers and Nanowires Synthesis Techniques

Processing techniques for nanofibers and nanowires fabrications and manufacturing broadly falls into two categories: top down and bottom up. Top down is defined as the breaking of large entities into smaller ones to nano- range. Bottom up is an assembly of small building blocks, atoms or molecules or even smaller things to design and fabricate hierarchies and complex systems by various wet chemical or physical methods. Nature stands behind the bottom up approaches as shapes, textures, sizes and designing of plant tissues, human cells and crystals all reflects the bottom up processing quite well. Bottom up approaches includes wet chemical methods like sol-gel, precipitation, co-precipitation, hydrothermal, solvothermal, template oriented, radiation based (sono chemical and microwaves), self-assembly, CVD (chemical vapor depositions), carbon-arc discharge, laser pyrolysis and much more to go. Template based synthesis, precipitation and sol-gel techniques provide good control on nanofiber dimensions in comparison to self-assembly and radiation based methods (Leung et al. 2011; Niraj 2005) (Table 4.1).

4.6.1 Template Based Growth

Bio-templating self-organized nano wires attracting a lot of attention for nano-micro scaled complex structures. Instead of typical and tedious physical and chemical processing techniques, the use of unique and interesting natural entities as template is now being explored. Template based nanotubes, nanowires (NWs), by using biological entities such as micro-organisms like fungus, bacteria, virus, spider silk, chitin fibers, collagen, cotton and bamboo etc., as soft templateis now in common practice (Sabah 2013). Among all alumina membranes are quite suitable to be used as hard template for fiber fabrication. Basically, template method is used to control the geometric features of nano-wires such as density and diameter by colloidal dispersion. The drawback islong time polarization effect and non-uniform pore filling for high aspect-ratio of nanorods and tubes cause template degeneration. Appropriate selection of template provide controlled and desired size, structure morphology, uniformity and density of the pores. Sol-gel technique is quite common for fabrication of well-ordered mesoporous metal oxidesfor sensing and energy applications by using surfactant as templates. The obtained nanofibers have uniform diameters within the range of roughly 25–120 nm (Zhao et al. 2007). Bio-oriented methods are simpler, versatile, environment-friendlyfor producing nanofibers and wires at industrial scales.

Table 4.1 Advantages and disadvantages of synthesis techniques for nanowires and nanofibers

Techniques	Advantages	Disadvantages
Template synthesis	Fibers of various shapes and sizes can be fabricated by using different template living or non-living	Hard (non-living) template are difficult to remove. A complex and especially designed process is required in many cases.
Phase separation	Minimum equipment requirement, direct fabrication of nanofibers matrix, mechanical properties can be controlled by varying the polymer concentration	Limited to types of specific polymers
Self-assembly	Good to get small nanofibers and composites without requiring any external glimpse	Complex process
Electro-spinning	Cost- effective, elongated and continuous nanofibers can be obtained	Jet instability
		Adjustment of voltage for spinning.
		Still hallmark to fabricate 3-dimensional structure
Precipitation and sol-gel	Easy to handle for fiber fabrication	Difficult to obtain excellent fibers without controlled conditions on large scale
Template free polymerization	Pure and uniform poly-nano fibers in large quantity.	Handling and choice of suitable polymer is difficult.
Hydrothermal synthesis	Cost effective and controlled environment at suitable temperature and pressure.	Nano-fibers and Nanowires formed with low crystallinity
Vapor-liquid-solid growth followed by the chemical vapor deposition.	Uni-directional and hybrid nano-wires having no defects are obtained	Choice of catalyst, must be inert

4.6.2 Phase Separation

Porous polymer templateswere employed by thermally induced phase separation in old era that are suitable for implantation due to their porous structure and interconnected spaces. Suitable attachment and migration of cells through interconnecting mechanical strength by such continuous three-dimensionalfibrous network make it an effective approach. This method involves following steps (i) polymer **dissolution (synonymous),** (ii) phase separation and gelation, (iii) solvent extraction through water (iv) freezingand drying under vacuum. The proper solvent selection is critical step or the formation nanofibrous structure. Thesemicroporous fibers provide three in one action to cell, first is nutrient absorption, second, the signals receiving and transmission, finally the waste disposal. This combination of nano- and macro-structures at the nanofiber level provides uniform cellular distribution and better response.

4.6.3 Self-Assembly

Self-assembly is highly promising bottom-up approachof spontaneously assembled smaller molecules into well-organized nanothreads. Variety of interactions are involved in the formation of these nano-structures like hydrophobic forces, dipole/dipole interactions, non-specific van der Waals interactions p–p stacking, hydrogen bonds, electrostatic interactions, and repulsive steric forces. Peptide amphiphiles chains consist of hydrophobic tails (dialkyle chain) and hydrophilic head (N-a-amino) group scan be self-assembled reversibly into nanofibers and can be applied for variety of materials fabrication. Many reagents such as acid, divalent ion and covalent moieties are responsible for the self-assembling of peptides. It produced low poly disperse nanofibers with high yield. Such nano fibers can develop "smart" biomaterial scaffolds for effective tissue regeneration in future. Briefly, the physical functionality of nanoparticles combined with bio-functioning of living cells as scaffold without any involvement of external entity and intermolecular bond breaking or making, entropic contributions and small enthalpy (ΔH) also play the prominent role in such self-assembled process (Fredric 2002).

4.6.4 Electrospinning

In this process, electrostatic forces fabricate continuous nanofiber assemblies. By applying high voltage to the droplet in solution, the molecules get charged and create an electrostatic repulsion, which counteracts the surface tension of the droplet. Variety of fibrous architectures was created based on this principle as electrospinning setups and collectors etc. The morphology of the fibers was strongly influenced by the surface tension, viscosity, and density like properties. Poly lactic acid (PLA) and tetra ethyl benzyl ammonium chloride (TEBAC) are used to fabricate fibers having low diameter by using electrospinning technique. Surface tension and electrical conductivity of the solution is controlled by an amount of TEBAC. These electrospinning nanofibers can be used as drug delivery reservoirs for controlled and timely release of drugs, proteins, antioxidants, and other molecules to the malfunctioning area in tissues. Electrospinning technique has plenty of benefits but it is still a challenge to fabricate a three-dimensional (3D) scaffold by electrospinning. Nanofibers have great capacity for the attachment or release of functional groups, absorbed molecules, ions, catalysts, and nanoparticles due to remarkably high surface area. As a result, these nanofiber assemblies can be used as highly active bio-scaffolds for cells migration and adhesion in the field of tissue engineering and sensing. Nanofibers of different materials as prepared by electrospinning are displayed in Fig. 4.1.

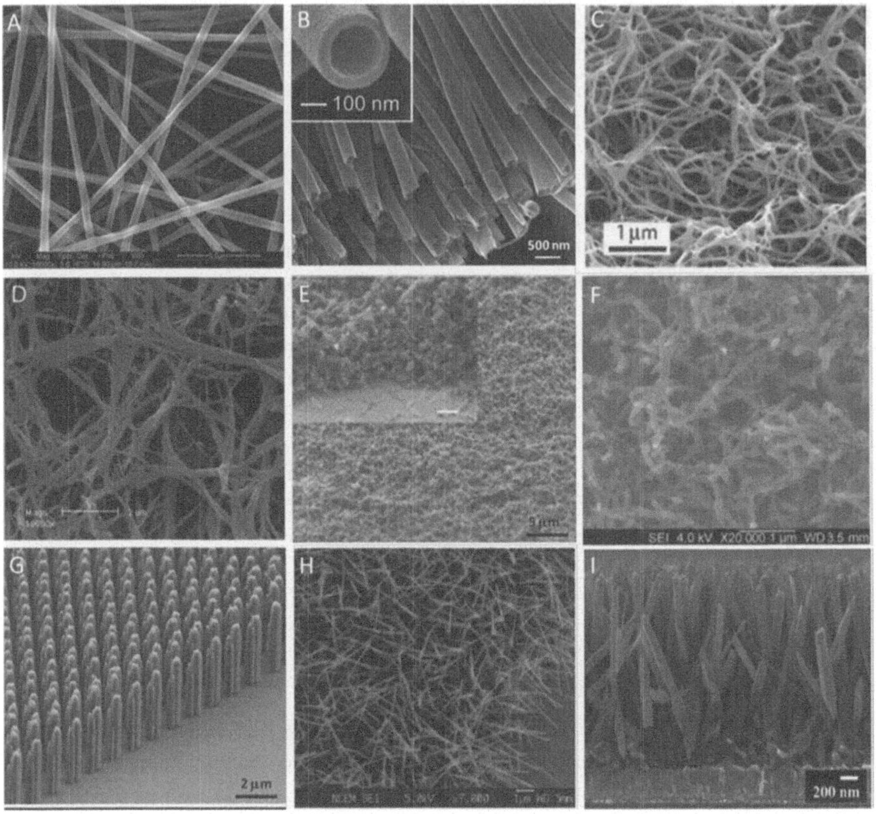

Fig. 4.1 SEM images of different types of nanofibers fabricated by electrospinning (**a**) random-oriented ribbons (**b**) aligned hollow tubes, (**c**) spider-web-shaped, (**d**) densed network -wires (**e**) rice grain-shaped (**f**) multi-channal entagled tubular (**g**) tubes in tubes (**h**) Needle -like wires mesh (**l**) vertically-aligned flattend ribbons (Jeong et al. 2015)

4.6.5 Template-Free (Interfacial Polymerization)

It is a template free technique to synthesize pure and uniform polyaniline nanofibers in bulk. Type and concentration of an aciddecide the dimensions, quality and uniformity of the nanofibers. High acid concentration results in the highproportion of nanofibers in the final product.

Rapidly initiated polymerization is an alternative technique to surface-interface polymerization (Abdolahi et al. 2012). Generally, an initiator solution containing ammonium peroxydisulfate in HCl is poured into a monomer solution containing aniline in HCl all at once. The size of the polyaniline nanofibers formed is influenced by the choice of an acid used, whilethe quality of nano-wires is effected the polarity of the reaction medium; more polar the solvent, more good the nanofibers.

Reaction temperature and reactant concentration have no effect on nanofibers morphology of nano-wires is not affected by temperature and concentration of reactants.

4.6.6 *Vapor-Liquid-Solid (VLS) Growth*

One-dimensional (1D) nanostructures without defects can be synthesized by this technique. Growth of hybrid and doped form of nano-wires and nano-fibers based on the formation of metal nanodroplets from gaseous precursors is crucial pint for this method. At first the metallic nanocrystals are melted in a tube furnace on a single stub. Saturation of molten metal nano droplets occur during melting process by introducing the different gases. These droplets act as catalysts to synthesize single crystalline and uni-directional nano-fibers. The interface energy plays an important role for deciding the droplet size and shape. If contact angle between solid surface and droplet results in larger growth area whereas, small contact angles give wires and whiskers of small radius.

4.6.7 *Hydrothermal Synthesis*

In this technique, the growth of the nanofibers and nanowires occur by chemical decomposition of thermally agitated reactive ions of precursors; in an autoclave under pressure of 1 atm and temperature of about 100–300 °C. Appropriate catalysts are used to enhanced the growth rate. Temperature, precursor concentration, and pH, etc. are the key parameters to control the fibers growth. Low level of crystalline structure of the grown nanofibers is one of the major drawback of this method. Among various nanofibers fabrication methods like self-assembly, phase separation, hydrothermal template based etc., electrospinning is the solely method to fabricate 1D ceramic nanofibers in a cost-effective and controlled environment.

4.6.8 *Sol-Gel Synthesis*

Sol-gel is one of the simple and flexible technique for 1-D nanowires synthesis. Initially the chemical solution as sol is used as precursor to precede final consolidated gel like network of either individual particles or polymers. Homogeneous polymer-embedded cellulose nanofiber composites were obtained by sol-gel process (Fig. 4.2). Cellulose whiskers are produced by the formation of homogeneous dispersed whisker as sol and thus followed by gel via exchange of water molecules with solvent. Acetone was routinely used in the procedure. Gels with a whisker

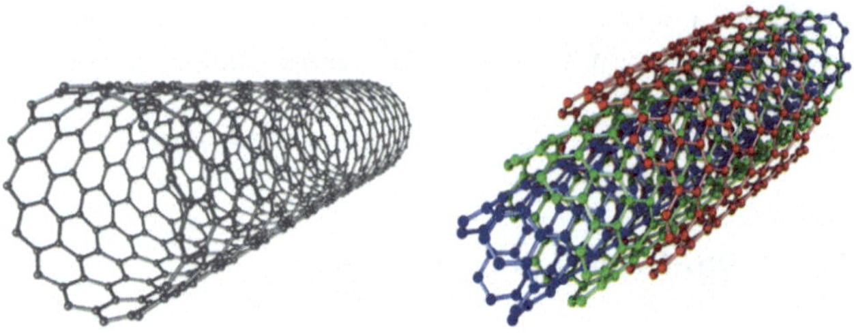

Fig. 4.2 (**a**) Single –walled and (**b**) Multi-walled carbon nanotubes

content of 1.0–1.8% w/w were explored. This method is widely applicable for the fabrication of nanocomposites having unique properties of individual components. These 'whiskers' exhibits high mechanical strength and can be obtained from naturally occurring bio-entities like wood and cotton, the safe and cost- effective (Jeffery et al. 2007).

4.7 Applications of Nanofibers

4.7.1 Carbon Nanotubes

As implantable sensors: Carbon nanotubes and their allotropes based sensors are frequently used as sensors for nanomedicineand nano-biotechnology due to their unique and enhanced properties. Properties such as surface area to mass ratio, retention time within tissues and body and the reactivity have greater impact in fuel cells, plasmonics, photovoltaic batteries, medical monitoring and diagnostics (Wujcik and Monty 2013). SWCNTs have a typical diameter of 0.5–1.5 nm, while that of MWCNTs are typically above 100 nm (Karimi et al. 2013).

CNTs are very promising for *in vitro* and *in vivo* sensing due to their unique properties such as an efficient entry into the cells, modification to get rid of toxic materials, enhanced intrinsic electronic, optical and magnetic properties that can be tuned and pertinent to life science applications. MWNTs are found to be encouraging scaffolds for orthopedic implantation, as well as provide skillful monitoring of bone growth through electrochemical sensing. Highly conductive surface area of MWNT's provides high electric activity in orthopedic treatment. CNT's can be prepared mainly by chemical vapor deposition (CVD), laser ablation etc. A remarkable surface area of SWNT's is promising for single molecule detection due to NIR fluorescence, band gap sensitivity to local environment and the quenching capabilities of CNT's. DNA modified SWNT causing red shift in band gap spectrum to modulate the transitions in blood, mammalians cells and tissues. Multi-walled and single walled CNTs are shown in Fig. 4.3. DNA and RNA strands can be detected very

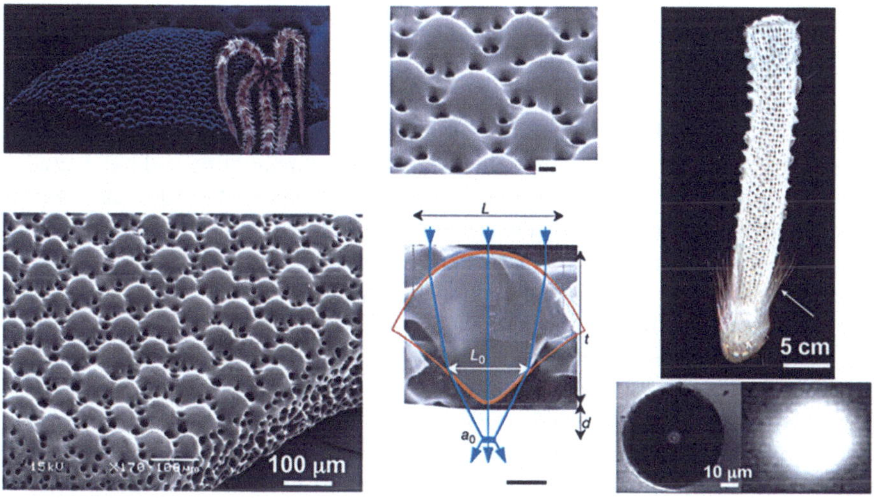

Fig. 4.3 Microlenses and fiber optical fabricated from protein scaffolds (Zhang 2003)

quickly by using CNT nano electrodes for molecular diagnosis. Well aligned multi-walled nano- tube arrays were embedded on SiO_2 matrix by bottom up approach.

In future, carbon based nano tubes can provide fast and appropriate detection of metal ions, contaminants neuron-transmitters and pathogen detection in microfluidic devices (Li et al. 2003).

4.7.2 *Difference in Carbon Nanotubes and Nanofibers*

CNT's and CNF can be differentiated to some extent as CNT's graphene layers are folded into hollow cylinders whereas in CNF, the graphene layers can be stacked as platelet, ribbon or rod like structure. CNFs can be termed as stacked-cup CNTs. The CNFs possess higher disorder degree than CNTs (the synthesis of CNTs carries out at higher temperatures). Difference can be examined by TEM and Raman analysis. The CNTs have lower diameter that CNFs. The diameter of CNFs can be higher than 80–100 nm. Tubes are the rolled graphene sheet. Fibers are all the carbon nano chips stacking on top of each other and form the fiber form. The diameter of CNF is larger than single-wall nanotubes.

4.7.3 Biological and Chemical Application of Nanofibers/Wires

Nanofibers enhanced the regeneration of many tissues and organs in medical applications for example, cell migration, infiltration, bio-compatibility and bio degradation with nano-fiber based scaffolds. Current challenges of tissues and organs failures can be accomplished by the use of nano fibers. High surface area to volume ratio of nano-fibers enhances the serum absorption by giving more binding sites to cells. Nanofibers also exhibit higher efficiency and flux in catalytic reactions (both hetro and homogeneous) chemical sensing and detection due to higher sensitivity in reactions in comparison to micro and nano-structures. Nano fibrous scaffolds facilitate skin regeneration, diabetic ulcer and normal wounds effectively in comparison to simple fibrous scaffolds. Nano-fibers are remarkably in use for investigating skeletal tissue engineering, both in absence and presence of growth factors. 3-D hierarchical nano-wires and fibers scaffolds are found to be mechanically strongwith controlled pore size and diameter for hard tissue regeneration (Nguyen et al. 2013).

ZnO nanorods grown on polyethylene fibers are used as antibacterial water purifier. Up to 99% of *E. coli* and *S. aureus* in spiked water containing about 1010 colony forming units (CFU) of bacterial cells were found to be immobilized under sunlight, while 80% of *E. coli* and 59% of *S. aureus* cells were inactivated. This water filter can be utilized in areas of drinking water scarcity. Formation of reactive oxygen contents through photo catalysis and release of Zn ions through dissolution of ZnO are responsible for antibacterial effect (Sunadan et al. 2012).

Living organisms at ambient temperature and low pressure can be used to fabricate composites. For example, fiber-optical spicule of Euplectella has the dimensions of a single human hair and can act as multimode waveguides. These optical fibers exhibit much more fracture resistance due to presence of organic ligands, high refractive indices and remarkable flexibility as compared with typical synthetic optical fibers. These microfibers act as multimode wave guides when light was coupled to free-standing spicules as depicted in Fig. 4.4 (Zhang 2003).

Silver nano-fibers were reported by using bio-organisms extracts as reducing and stabilizing agents. Single crystalline Ag nano plates can be synthesized by the extract of unicellular green algae Chlorella vulgaris at standard temperature pressure conditions. Proteins in the extract act as reducing agent for Ag ions as well as stabilizing agent for controlling the shape of fibers. Basically the carboxyl groups in aspartic and glutamine residual and the hydroxyl groups in tyrosin residue of the proteins were the cause of Ag + ion reduction. Extracts from live plant alfalfa, the liquid broths of lemongrass, geranium leaves etc. are used as green reactants in Ag NF synthesis. A mushroom, vegetable like capsicum etc. can also produce Ag-protein (core–shell) NPs with an average size of 30.5 nm. Glutathione (γ-Glu-Cys-Gly-), a bio-organic act as reducing and stabilizing agent produce water-soluble and size tunable Ag NFs that can attached to protein, bovine serum albumin, dominant in bio-nano-medicine (Sharma et al. 2009).

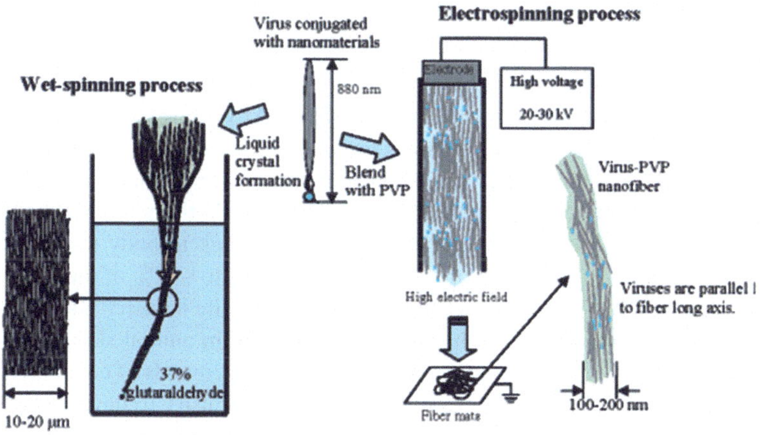

Fig. 4.4 A schematic representation of the virus-mimicking fiber fabrication process by wet-electro-spinning technique (Lee and Angela 2003)

M13 viruses were used to fabricate micro and nano sized fibers by following the electro and wet spinning techniques of the silk spideras depicted in Fig. 4.4. In capillary tubes (diameter ~micro-meter), virus suspension was properly shaped by cross linkage of glutaraldehyde solution. Spider-silk mimicking fibers were found to be 10–20 μm in diameter. After that mixing of polyvinyl pyrolidone (PVP) and M13 viruses produce continuous uniform virus blended PVP nano-fibers. These electro spun fibers are prominent in sensitive catalytic actions for biomedical field and biosensors in future (Lee and Angela 2003).

Polymer nano-fibers provided a proper route to a bio-mimetic approach. Surface features and topography display the important effects on to control cell adhesion, arrangement, orientation, activities etc. within all around nano and micro featured network. These polymer-based nano fibers and wires are used prominently in tissue engineering, drug attachment and delivery, wounds healings, medical implants, nano-composites for dental applications, molecular separation, biosensors and preservation of bioactive agents.

Actin and tubulin are naturally occurring filamentous protein structures can be used as template for nano and micro wires. The first use of such cytoskeleton proteins for building the nanostructure was the utilization of s7 nm thin actin filaments as templates for nanowires fabrication. Self-assembly of actin protein filaments was covalently modified by depositing gold nanoparticles using an amine-reactive agent (N-hydroxysuccinimide), results in continuous gold nano-wires. Actin fibers and ATP hydrolysis are used to give cytoskeleton modification that may lead to various nano-electromechanical system applications in mechanics, conductivity based on the bio-inorganic composites (Gazit 2006).

4.7.4 Metallic Nano-wires and Textile Industry

Metal-coated nano-wires embedded cloths are more durable for several washing cycles due to increased mechanical strength. Silver (Ag) deposited nano-wires coating in textile provides thermal insulation more than 21% due to less radiation loss. Briefly, the metallic nano-wires form a mesh of wires having empty spaces of 200–300 nm with nano-wires in neighborhood, so that visible light can be transmitted and majority of the human body radiations reflected back to body to enhance the thermal insulation. This spacing of 300 nm corresponds to ~97% of thermal reflectance (Hsu et al. 2014). It's almost 70 years ago, that textile industry made a revolutionary impact to our society with dramatic changes from animal skin to leaves to yarns and now fabrics. In order to meet with increasing demands of consumers, new technologies are needed to apply to the trendy textile industry to develop the products with remarkable values to overcome the issues regarding with conventional finishing treatments. Growth of oriented hexagonal ZnO nanoparticles rod arrays (10–50 nm in diameter and 300 nm to 500 nm in length) onto various flexible fiber substrates by using dip-coating and self-assembly processes is reported.

ZnO dumb-bells shaped nanostructures provide a high UPF (UV protection factor) of 1000 and demonstrated a breakthrough of more complete blocking UV range (400–280 nm) as compared of ZnO nano-sols (352–280 nm), nano-rods (375–280 nm), and anatase Titania films (332–280 nm). Such unique shaped ZnO based architecture is not only prominent in textile field as well as showed great implication in vehicles, windscreens, glass windows, outdoor products, and aircrafts industry. A fabric functionalized by Alkyl silanes modified silica on oriented growth of ZnO crystals displayed increases water durability. The water contact angle was decreased to 128°, 120°, 114°, and 110°. Basically, the GPTS and F13 promoted the adhesive forces to cloth resulting in durable wetting even after repeated washing cycles. Finally, the chitosan fibers as core material (~50 nm to 5 μm) and gold shells as capped entity (~20–50 nm in depth) were used for fabricating organic and inorganic complexes (Wang 2007).

4.7.5 Nanowires and Nanofibers for Sensing Applications

Immobilization of antibodies and bacterial binding on nano-diamond and carbon nano-tubes is crucial for biosensor applications. Biosensors are the devices used to detect the biological changes in human body. High surface area, flexibility, increases mechanical strength and elasticity, and specifically directed growth, made the fiber a suitable material for various applications ranging from textile industry to air-crafts. One-dimensional nano-wires displayed great capacity for the attachment or release of functional groups, absorbed molecules, ions, catalysts, and nano-particles due to its remarkable increase in surface area per unit volume that is highly demanding for sensing and cell engineering as well (Wang 2006).

Sensors based on silicon nano-bio-wires SiNW, nano-arrays and complexes increases light absorption, charge transfer and separation, chemical and catalytic performances. SiNW FETs are used in biological interfacing with live cells and tissues for intracellular electrical measurements, and electrical diagnostic of biological and chemical species (Yanli et al. 2013).

References

Abdolahi A, Esah H, Zaharah I, Shahrir H (2012) Synthesis of uniform polyaniline nanofibers through interfacial polymerization. Materials 5:1487–1494

Alexandra B, Gow NAR (2009) Mechanisms of hypha orientation of fungi. Curr Opi Microbiol 12(4):350–357

Bae J, Jin-Ho C, Yong SY, Nam KO, Byung SK, Myongsoo L (2005) Helical nanofibers from aqueous self-assembly of an oligo (*p*-phenylene)-based molecular dumbbell. J Am Chem Soc 127(27):9668–9669

Cassie ABD, Baxter S (1994) Wettability of the porous structures. Trans Faraday Soc 40:546–551

Chang J, Xiu-Feng P, Karam H, Joseph C, Hamidreza G, Santiago DS, Joonil S (2011) Nanomechanical stimulus accelerates and directs the self-assembly of silk-elastin-like nanofibers, communication. J Am Chem Soc 133(6):1745–1747

Chappel S, Zaban A (2002) Nanoporous SnO_2 electrodes for dye-sensitized solar cells: Improved cell performance by the synthesis of 18 nm SnO_2 colloids. Solar Energy Mat Solar Cells 71:141–152

Fredric MM (2002) Supramolecular chemistry and self-assembly. Proc National Acad Sci 99(8):4818–4822

Gabor LH, Tibbals HF, Joydeep D, John JM (2008) Introduction of nano science and nanotechnology, chapter no: 3, December 22. CRC Press, Boca Raton

Gazit E (2006) Use of biomolecular templates for fabrication of metal nanowires. The FEBS J 274(2):317–322

Gow ANR (1994) Growth and guidance of the fungal hypha. Microbiol 140(12):3193

Hornyak GL (1997) Characterization and optical theory of nanometal/porous composite membranes, PhD dissertation, Colorado State University

Hsu PH, Xiaoge L, Chong L, Xing X, Hye RL, Alex JW, Tom Z, Yi C (2014) Personal thermal management by metallic nanowire-coated textile. Nano Lett 15(1):365–371

Huang F, Wei Q, Cai Y (2012) Surface functionalization of polymer nanofibers. A volume in Woodhead Publishing Series in Textiles:92–118

Jeffery RC, Otto VDB, Lynn AC, Michael S, Stuart JR, Dustin JT, Christopher W (2007) A versatile approach for the processing of polymer nanocomposites with self- assembled nanofiber templates. Nat Nanotech 2:765–769

Jeong F, Kim, JH, Kim, Y, Moo, L, Enrico D (2015). Thermally-induced phase separation (tips) and electrospinning methods for emerging membrane applications: a review. AICh E J 62(2):461–490

Kamps AC, Michael F, So-Jung P (2012) Hierarchical self-assembly of amphiphilic semiconducting polymers into isolated, bundled, and branched nanofibers. ACS Nano 6(3):2844–2852

Karimi M et al (2013) Carbon nanotubes part I: preparation of a novel and versatile drug-delivery vehicle. Expert Opin Drug Deliv 12(7):1071–1087

Lee SW, Angela MB (2003) Virus-based fabrication of micro and nanofibers using electrospinning. Nano Lett 4(3):387–390

Leung KM, Wanger G, Guo Q, Yuri G, Gordon S, Woon ML, Jun Y (2011) Bacterial nanowires: conductive as silicon, soft as polymer. Soft Matt 7:6617–6621
Li D, Xia Y (2004) Direct Fabrication of Composite and Ceramic Hollow Nanofibers by Electrospinning. Nano Lett 4(5):933–938
Li J, Hou TN, Alan C, Wendy F, Hua C, Qi Y, Jessica K, Jie H, Meyyappan M (2003) Carbon nanotube nanoelectrode array as an electronic chip for ultrasensitive label-free DNA detection. Nano Lett 5:597–602
Malhotra BD, Saurabh S, Shine A (2015) Biosensors for food toxin detection: carbon nanotubes and graphene, MRS Online Proceeding Library Archive:1725. https://www.researchgate.net/publication/276394214
McCann JT, Li D, Younan X (2005) Electrospinning of nanofibers with core-sheath, hollow, or porous structures. J Mater Chem 15:735–738
Nguyen LTH, Shilin C, Naveen KE, Molamma PP, Zong Y, Chellappan V, Suleyman IA, Seeram R (2013) Biological, chemical, and electronic applications of nanofibers. Macromol Mat Eng 298(8):822–867
Niraj S (2005) Carbon nanotubes for biomedical applications. IEEE Trans Nanobiosci 4(2):180–195
Pérez-Juste J, Pastoriza SI, Liz-Marzán LM, Mulvaney P (2005) Gold nanorods: Characterization and applications. Coord Chem Rev 249:1870–1901
Sabah A (2013) Self-organization of charge stabilized gold nanoparticles on directed fungal mycelia directed by external stimuli, PhD dissertation, Asian Institute of Technology AIT, Thailand
Sharma VK, Ria AY, Lin Y (2009) Silver nanoparticles green synthesis and their antimicrobial activities. Adv Coll Interf Sci 145:83–96
Sunandan B, Samir KP, Joydeep D (2012) Nanostructured zinc oxide for water treatment. Nanosci Nanotech-Asia 2(2):90–102
Tao XY, Zhang XB, Zhang L, Cheng JP, Liu F, Luo JH, Luo ZQ, Geise HJ (2006) Multi-branched and porous carbon nano-fibers were prepared for efficient performance of double layer capacitors. Carbon 44(8):1425–1428
Wang R (2007) Multi-functional protective nano-finishing of textile materials. The Hong Kong Polytechnic University. http://ira.lib.polyu.edu.hk/handle/10397/2277
Wang Y, Tianyu W, Peimei D, Ming X, Hao W, Gengfeng Z (2013) Silicon nanowires for biosensing, energy storage, and conversion. Adv Mat 25(37):5177–5195
Wu A, Heidelberg A, Boland JJ (2005) Mechanical properties of ultrahigh-strength gold nanowires. Nat Mat 4:525–529
Wujcik EK, Monty CN (2013) Nanotechnology for implantable sensors: carbon nanotubes and graphene in medicine. WIREs. Nanomed Nanobiotechnol 5(3):233–249
Yury G (2006) Nanotubes and nanofibers. CRC Press/Taylor & Francis Publishing Group, London
Zhang S (2003) Fabrication of novel biomaterials through molecular self-assembly. Focus Nanotech 121:1171–1178
Zhao HP, Xi QF, Huajian G (2007) Ultrasonic technique for extracting nanofibers from nature materials. Appl Phys Lett 90:073112

Chapter 5
Nano Drugs

Aneeqa Sabah and Gabour Loius Hornyak

5.1 Nano Drugs, an Outlook

Nanotechnology is an upcoming interdisciplinary approach with vast applications in biology, chemistry, medicine together with networking and communication technology. A number of devices are fabricated at miniature level based on bio-mimicking nano hierarchical structures. Further working these nano-devices is controlled and analyzed by numerous equipment and techniques at nano-scale. Novel properties, variety of responses and functions of living and non-living species at nano-scale are investigated and modified by using these techniques and tools. The potential medical applications are predominantly in detection, diagnostics (disease diagnosis and imaging), monitoring, and therapeutics. Recent advances depicted that drug-delivery systems are of great scientific interest and give better hope for the treatment of cancer, heart disease, diabetes and much more. Many novel nano-particles, nanostructures and nano devices are expected to be used, with an enormous positive impact on human health. The vision is to improve health by enhancing the efficacy and safety of nano systems and devices. Materials either chemical or biological can be produced with high achievements and interesting properties rather than traditional synthetic techniques. In future, nano-particles are expected to be used in drug- attachment and targeting systems. Due to small size, nano-particles cannot be identified within the human body and easily travel through cell membranes beneath a critical size and are able to move through the blood - brain barrier without creating serious hazardous.

A. Sabah (✉)
Department of Physics, Lahore College for Women University, Lahore, Pakistan

G. L. Hornyak
Asian Institute of Technology, Khlong Nueng, Thailand

S. Javad, A. Butt (eds.), *Nanobotany*, https://doi.org/10.1007/978-3-319-77119-9_5

It is very difficult for diabetic patient to inject insulin many times a day or for cancer patient to go through many side effects of treatment. Hence the drug delivery system may change their life to easy way of taking medicine.

It is almost 100 years ago that nana-objects are in use for variety of field but nano medicine is established as remote interdisciplinary science in twenty-first century. The key of nano-medicine development is based on the knowledge and detailed investigation of cellular, molecular structures in biology, chemistry and physics. The nano technological approach has shown tremendous and rapid development in the field of medicine to meet modern requirements, for examples tissue engineering and cells fictionalization and treatment. Many bio-compatible nano materials are in the form of drug, vaccine and genes for different therapies. Bio-sensors are frequently used to diagnose the diseases by many techniques like nano-imagining and lab-on-chip. Nano-capsules may found to suitable for cancer treatment, hyperthermia etc. Stent angioplasty, regenerative medicine and gene therapy are becoming more prominent now days.

5.2 Nano Drug Routes/Drug Injection Routes

There are many other ways to deliver the drug more efficiently and less unpleasant for the patient. Some are just in the process of examined but some are already being tested and actually being in use today. Drug can be taken orally, in rectal and topical way, may be injected and inhaled. Nanoparticles are categorized to be used in drug and gene delivery and also to prob. DNA strands such as polymer-capped nano capsules and nano spheres, lipid-based liposomes, crystalline nano-structures, dendrimers, fullerens C60 (Hornyak et al. 2008), metallic nanoparticles (e.g. gold Au and silver Ag) as well as magnetic nanostructures.

5.2.1 Oral Drug Delivery

The oral route is simple and frequently used pathway for drug delivery. It is more convenient and less painful for the patients. Self- application for oral delivery results in high compliance. The intestinal path provides a large surface area for absorption of 300–400 m^2), an excellent blood supply, and extensive residence time, good for drug absorption with good solubility and permeability (Sanganalmath and Bolli 2013; Fukushima et al. 2007). However, most drugs met with problems of stability, permeability or solubility in the gastro-intestinal tract area so the result is an inappropriate intra and inter movement of drugs, low bioavailability, unpredictable absorption and disturbed proportion of required dose. Many new therapeutic drugs cannot be developed for use as conventional oral formulations, due to the inhospitable environment found in the gastrointestinal tract, such as high metabolic activity and pH variations.

Limitations associated with oral drug delivery can be overcome by using nano-based encapsulation of drugs, peptides and proteins in nano-polymers improves the water solubility to enhance the gastrointestinal permeability for drug targeting without effected by local pH and enzymes.

5.2.2 *Intranasal Drug Delivery*

Advancement in nanotechnology has enabled the nasal route to be involved in anticancer drugs, analgesics, central nervous system (CNS) drugs, peptides, and diagnostic agents. The nasal mucosa is suitable for the systemic absorption of drugs due to epithelial microvilli, large surface area, rich vasculature, and a highly porous endothelial membrane) to facilitate drug permeation, such that drugs can be absorbed directly into the systemic circulation without the first pass effect (Shen et al. 2013; Molina et al. 2009). Dose quantity and side effects are reduced dose reduction, whereas, patient wish for treatment increases due to all above described effects. Additionally, the intranasal administration provides a non-invasive and painless alternative to the intravenous and oral routes to maximize the patient comfort. Beside all these advantages, there are some limitations includes mucociliary clearance, that rapidly removes the formulation from the nasal cavity. In addition, enzymatic degradation occur in both the lumen of the nasal cavity and passing through the epithelial barrier (Molina et al. 2008). Other factors are the low permeability of the epithelium that hinder the transference of polar drugs or high molecular weight substances such as peptides and proteins, the pH of the system must be compatible with the nasal cavity (Sun et al. 2014) and the small volume that can be administered and the mucus layer. Polymeric nanoparticles show many advantages as systems for nasal administration, for example, alteration of the mucus layer, tight junctions, erosion of the mucosal surface, increased drug contact time at the absorption site, the use of bio-adhesive materials.

Systematic delivery of peptides and proteins may possible by nasal drug-delivery methods. Their instability in acid medium, high molecular weight, and hydrophilicity make peptides and proteins travelling difficult through the gut epithelium by oral- system. Further, the general administration of these macromolecules involves the parenteral route, generating drawbacks such as the requirement for a qualified professional for administration, patient compliance, and high production cost. Thus, the nasal route increases the bioavailability of these agents. Insulin-loaded nanoparticles administered by the nasal route increases the biological half-life and improve the stability and therapeutic efficiency for protein delivery by using many polymers (Xie et al. 2012; Lencioni and Bartotozzi 1996; Britten et al. 2003).

5.2.3 Dermal and Transdermal Drug Delivery

Primarily, the nanoparticle medicine-delivery system is focused on parenteral and oral applications. Nowadays, nano systems may also applied to the skin diseases. Nano-entities are used as protective agent against chemical decay and flexibility to regulate the release of elements in dermal field (Zhang et al. 2014a). Nano-treatment to the skin may be divided into two aspects: 1: dermal drug delivery within the skin or systemic activity by nanoparticle permeation through the skin, i.e. trans-dermal drug delivery (Kutty and Feng 2013; Wang et al. 2011).

In both the dermal and trans-dermal routes, the horny outer epidermal layer is the main hinderance of the skin, has to be overcome for suitable drug delivery (Kutty and Feng 2013). The skin is composed of the epidermis, dermis, and subcutaneous tissue. The epidermis is further subdivided into four layers, the strata corneum, granulosum, germinativum, and basales. The cells in the stratum basale divide continuously to produce new keratinocytes that move to the outer layers and form the stratum corneum, a horny layer of dead cells. In addition, the adaptable epidermis contains cells such as melanocytes, sensory intuition Merkel cells, immunological function as Langerhans and the sub-ordinates of plants etc. The main role of the skin is to protect from physical, immunological, metabolic, and UV-irradiation toxics, microbes, hazardous chemicals (Zhang et al. 2014b). However, drug can easily pass through skin layers due to high surface area and easy exposure. Three main penetration pathways can be identified in the skin. 1: Intracellular across the corneocytes, intercellular by the lipid bilayers and trans appendicular, includes hair follicles and sweating gland routes. Among all, the intercellular route is found to be feasible and best pathway. In fact, dominance amongst the three routes depends on the drug type, its solubility, diffuse ability, molecular size and physico-chemical properties and also the system used for delivery of the substance (Sajomsang et al. 2014).

5.2.4 Parental Drug Delivery

Generally, the parental pathways require low doses of drug to achieve the same effect as high doses with oral administration of drugs. This is mainly because the parenteral pathway allows for the rapid and complete absorption of drugs, unlike an oral route where a large fraction of the dose is lost due to liver metabolism. Furthermore, the parenteral route has extended therapeutic effects in comparison to other routes. Thus, reduce the side effects as compared to oral administration. Drugs with poor dissolution properties are good candidates of parenteral administration. Traditional techniques have so many issues like poor-biocompatibility, accurate dose proportionality and slow process, patient may lose interest. Continuous infusion of drugs with a short half-life as well as drug administration to unconscious patients is possible. Cardiovascular agents, anticancer, anti-inflammatory, antibiotic, immunomodulatory, and immunostimulatory drugs, antiglaucoma compounds, and even peptides can be entrapped in nanoparticles proposed for parenteral route (Na et al. 2010; Baranello et al. 2014; Ramasamy et al. 2014; Shi et al. 2014).

Nanoparticles administered by the parenteral route can be quickly captured by the mononuclear phagocyte system (MPS). In this case, blood cells such as monocytes, leukocytes, and platelets, as well as resident phagocytes such as (Baranello et al. 2014), Kupffer cells of the liver and spleen macrophages, play an effective role in drug-delivery process (Cho and Kwon 2014; Cho et al. 2013; Han et al. 2012).

5.3 Nano Drug-Delivery Systems

These systems are especially designed to overcome the drawback associated with conventional application of drugs like poor bio-distribution, stability, solubility, therapeutic effectiveness etc. (Emeje et al. 2012; Wilczemska et al. 2012; Haley and Frenkel 2008). In drug delivery systems, drug is being targeted directly on the effective area, avoiding the undesirable side effect. Even low dose of drug is enough to work properly.

Nanoparticle within the size range of 1–100 nm can be used as drug carriers having optimized physical, chemical and biological properties like small size, increased drug accumulation, ability to travel within the cell or tissue barriers, controlled drug release, etc. The use of nanoparticles as drug carriers plays an important role in eliminating the existing challenging problems due to conventional drug usage.

Polymeric micelles, liposomes, dendrimers, nanosphers, nanocapsules, nanotubes, nanogels and metallic nanoparticles like gold and silver contribute as nano-drug delivery systems (Estanqueiro et al. 2015; Dikmen et al. 2011; Guney et al. 2011).

Cancer treatment is secured and approved medically by the Food and Drug Administration (FDA). The cure process is divided into two parts: active and passive targeting. In active one: drug is modified by antigen or antibody and follows the polymer sensitive to variation in temperature, whereas, in passive treatment: increased accumulation and retention of drug on the targeted cell or tissues is required (Yhee et al. 2013).

5.3.1 Polymeric Micelles

Polymer based micelles are the amphiphilic co-polymers are designed specifically with inner and outer core or shell structures to deliver the drug on target area. Hydrophilic shell composed of hydrophilic non-biodegradable polymers, for example, poly(ethylene oxide) (PEO), poly(N-isopropylacrylamide) (PNIPA), poly (alkylacrylic acid), and poly(ethylene glycol) (PEG)). These polymers maintain the stability in an aqueous mediumand thus enable the interactions with proteins and cell membranes. The hydrophobic core capped various non-polar drugs, such as paclitaxel, doxorubicin, tamoxifen, camptothecin, porphyrins, etc. The hydrophobic

Table 5.1 Advantages and disadvantages of various drug-delivery systems

Advantages	Disadvantages
1: Polymeric micelles	
Small size 10 nm –100 nm	Difficult polymer synthesis
High structural stability- self-assembled lipid core micelles	Unstable- self-assembled polymeric micelles
Higher concentration of drug loading and sustained drug release	No universal in-corporation method
High water solubility for hydrophobic drugs	Liver chronic toxicity-slow metabolic process
Low toxicity	Toxic side effect -longer impact time
Modified by various chemical specie	Not applicable at large industrial scale
Protection of encapsulated drugs from degradation and metabolism	Only few polymeric micelles are approved by FDA for clinical use
2: Dendrimers	
Monodisperse hierarchical structures with good control than linear polymers	3-dimensional complex structure, so the secondary-structural motifs within the branches are difficult to understand
Terminal functional group have positive, negative and charges and neutral one	Non-specific interaction with the variety of cells and toxic effects.
3. Liposomes	
Liposome occurs with increased efficacy and therapeutic effects of drug	Production cost is high
Increases drug stability	Leakage and fusion of trapped drug molecule
Bio-compatible, bio-degradable, non-toxic, flexible and non-immunogenic liposomes	Degradation of phospholipid may occur sometimes
Reduction in toxicity of entrapped agent	Physico-chemical instability
Passive and selective active targeting to the tumor tissues	Low solubility and limited entrapping efficiency
Generally used in diagnosis and gene delivery	Unsterlization
4: SLN's solid-lipid nanoparticles	
Controlled drug release, low- term stability	Poor drug loading
Low toxicity of drugs and compounds	Degradation of drug during storage
Enhancing the bio-availability of entrapped drug	High water content
Chemical protection	Low capacity to load hydrophilic drugs

core consists of water-soluble polymers such as poly (propylene oxide) (PPO), poly(β-benzyl-L-aspartate) (PBLA) [30], poly (L-lactic acid) (PLLA), poly (lactic -co-glycolic acid) (PLGA), poly(ε-caprolactone) (PCL) or poly (aspartic acid) (PASP), polystyrene (PST), poly (methyl methacrylate) (PMMA), and also the polyacrylates [18–28]. Polymeric micelles occur withinsize range from 10–100 nm. Narrow size range of polymeric micelles ensures high stability, sterility and long term circulation in the bloodstream. Stability of polymer micelles is dependent on the polymer chain in inner core (Cho et al. 2014; Kumar et al. 2012; Yokoyama 2010). Only few polymeric micelles are approved by FDA and used in clinical applications. Table 5.1 displays advantages and disadvantages in various drug delivery systems (Eskiler et al. 2013).

Drug production and incorporation into polymeric micelles may be direct or indirect, based on drug quantity and polymer type. In direct approach, amphiphiles are dissolved in water and finally enclosed by the drug, the indirect method depends on the use of water-miscible organic solvents (e.g., acetone, dimethylacetamide) to solubilize the co-polymer and the drug and organic solvents are separated by evaporation or dialysis (Sosnik and Raskin 2015; Venkataraman et al. 2011; Gaucher et al. 2010; Lee et al. 2007). Drugs can be loaded into micelles both by chemically or physically through dialysis or emulsification method. Chemical embedding mayenable the drug to be incorporated with the hydrophobic polymer through covalent bond, Thus the encapsulated drug is protected from enzymatic cleavage due to non-hydrolysis of chemical bonds (Krishnamurthy et al. 2014).

Soluble and non-toxic solvents e.g., ethanol, N-N-dimethylformamide are used in the dialysis method, water-insoluble solvents e.g., chloroform, chlorinated are used in the oil -in-water emulsion method (Satchi-Fainaro et al. 2006; Katayose and Kataoka 1998; Weissig et al. 1998).

Micelles can be characterized by turbidity measurements, CMC critical micelle concentration and aggregate size. Micelle sizes can be settled by dynamic light scattering DLS technique (Kozlov et al. 2011; Schwarz et al. 1998; Nakamura et al. 1976).

5.3.2 *Solid Lipid Nanoparticles (SLN's)*

SLN's within size range of 100–1000 nm are good colloidal drug carriers to overcome the poor drug loading, size problem and unstable properties. Uncontrolled amount of drug release on effected area associated with liposomes, dendrimers and polymer micelles can be well controlled by using SLN's. They exhibit interesting properties like low toxicity, higher efficiency for large quantity of drug loading due to high surface area, controlled drug release and protection from degradation. SLN's used for delivering of lipophilic and hydrophilic drugs on target area. Recently, some attention is being paid for lipophobic drugs in gene delivery.

Preparation methods for SLN's includes high pressure homogenization (hot and cold homogenization), micro-emulsion, solvent emulsion diffusion, solvent evaporation, ultrasonication/high speed homogenization, and spray drying methods Mussi and Torchilin 2013; Patil et al. 2013; Mader and Mehnert 2004).

5.3.3 *Liposomes*

Liposomes are the closed spherical vesicles, made of natural and synthetic phospholipid bi-layer for encapsulation of drugs and genetic stuff. Hydrophilic drugs capped in a core material and hydrophobic will be trapped in lipid bi-layer. By liposomes, drugs can be delivered into cells by fusion or endocytosis mechanisms to provide controlled drug release and protected from degradation by plasma enzymes to reduce the side effects. Liposomes can be functionalized easily by coating the surface or attaching different specific surface biomarkers PEG, antibodies, and peptides to reach cancer cells. The properties of liposome such as size, permeability and charge density etc. can be determined by phospholipids constituents (Gupta et al. 2012). Liposomes can unilameller (composed of single lipid bi-layer) or multilamellar (several lipid bi-layer) ~ of 100 nm to 20 μm. Unilameller are further divided into two categories, 1: Small size unilameller 25–100 nm or large size unilameller 100–1000 nm (Lopes et al. 2013).

Several preparation methods having good control on properties are designed for liposomes such as size, structure, density of bi-layers and also the drug type and concentration. Synthesis medium also matters. Main three synthesis methods are described in literature 1: mechanical agitation, 2: solvent dispersion and 3: detergent removal method. They are further subdivided as sonication, ether injection, freeze-thawed, micro-emulsification and fluidization, spray drying, reverse-phase evaporation vesical, dialysis and dilution, layophylization etc (Laouini et al. 2012; Lopes et al. 2013; Jaafar-Maalej et al. 2010).

Liposomes can be characterized both qualitatively and quantitatively by observing the visual appearance and measurement, size distribution, surface-charge content on phospholipid, lamellarity, stability and efficiency (Makino and Shibata 2006; Ramandeep Singh 2016; Vemuri and Rhodes 1995).

5.3.4 Dendrimers

Dendrimers are unique, 3- dimensional structures, can be linear, cross linked and branched. They are hyperbranched globular-shaped particles. Dendrimers are specifically used for cancer treatment in drug and gene delivery system. Three main components comprise the dendrimers. 1: central core determines the size and shape of the dendrimer, 2: Tree like, star shaped monodispersed branched structure, 3: terminal functional group, located on exterior surface of dendrimers. Dendrimers are the good carriers of drug delivery and increases the therapeutic effectiveness and enable the targeting of any drug to corresponding site (Kesharwani et al. 2014; Tripathy and Das 2013).

5.3.4.1 Types of Dendrimers

To overcome limitations and improvement of application in delivery systems, different dendrimers types are developed with enhanced functionalities. PAMAM and PPI dendrimers, synthesized by divergent method, frequently used in gene transfection and having numerous applications in material science as well as biology. Liquid crystalline dendrimers (LC) shows limited applications in DNA delivery. Core

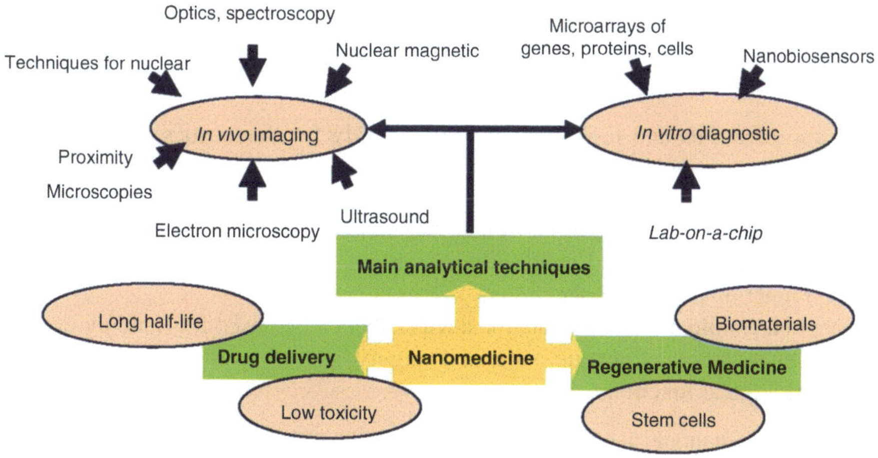

Fig. 5.1 Schematic representation of analytical techniques for diagnostics and drug delivery (Boulaiz et al. 2011)

shellare frequently used in recognition anddiagnosis and also the drug delivery. Chiral dendrimers act as chiral hosts for enantiomeric resolutions and a catalyst in asymmetric synthesis. Peptide produced by convergent and divergent methods used as surfactant and multiple antigen peptide in nanomedicine, protein holders and carriers for gene and drug delivery on industrial scale. Glyco-dendrimers holds sugar entities like glucose, galactose, maltose etc. and provide site-specific drug delivery in lectin rich organs, and hybrid dendrimers obtained from polymers generated the compact, rigid uniformly shaped to be used in drug delivering (Nasibullah et al. 2013; Singh and Sharma 2013; Tripathy and Das 2013; Welch and Welch 2009; Darbre and Reymond 2006; Dufes et al. 2005).

Dendrimers can be produced by two techniques. 1: The divergent technique is costly and difficult to handle, dendrimers grows outward from inner core shell. During the production process, some branches many not be extended smoothly. These deformations may include some impurities and also affect the functionality of dendrimers. 2. In convergent method, dendrimers are produced by functional molecules. Branches grow toward inner core, and eventually attached to the core. Impurities can be removed easily, but large quantities of dendrimers can't be produced by convergent method (Malik et al. 2012). The polymers used in dendrimer production are: PAMAM, poly(propylamine) (PPI), poly(L-lysine), triazine, melamine, PEG and carbohydrate-based citric acid, poly(glycerol-co-succinic acid), polyglycerol, and poly[2,2-bis(hydroxymethyl)propionic acid] (Carlmark et al. 2009). Dendrimers displays monodispersed or polydispersed frameworks. Dendrimers form tightly packed ball like structure after reaching the maximum size (Fig. 5.1).

5.4 Nanotechnology and Drug Delivery

In medicine, nanotechnology is categorized as follows:

1. Nano materials and tools are used prominently for diagnostics, treatments of diseases transportation of active substances through the living cells and analysis.
2. Know how of medicine at molecular and atomic level in the fields of genetics, protein interactions and analysis within living beings, synthetically produced and functionalized microbes.
3. Nano technologies to fabricate the devices that can be used for rapid diagnosis, therapies, fixing of genetic materials, surgery, as well as to ameliorate natural physiological functions.

The use of nano-structures to differentiate various types of issues and cells by *in vivo* imaging with an improved three-dimensional view is preferable. Quantum dots provide clear and focused visibility of genes, nucleic acids, proteins, molecules and cell processes. Due to the immense stability and brightness these photo probes display luminosity 1000 times greater than that of typical contrast media. The in-time presence, position and size of tumors can be diagnose and cures more accurately as compared to conventional methods and tools. Early and appropriate detection of defected areas, genetic displacements and diseases by using low-cost, rapid analysis and accurate nano-imaging techniques rising the expectation and believe of mankind on nano-medicine.

Development medicines free from side effects, close entangling of diagnosis and therapy, bone culture, tissue and organs with the help of nano-particles and adult stem cells will hopefully improve quality of life Bradbury 2003).

Now a day's very small amount of specimens are needed to be tested for measurements and diagnosis due to inventions of nano-sized portable and centralized equipments. These miniaturized lab-on-a-chip devices give precise measurements and burden-less environment to the patient and doctors. Such an easy access is very helpful in preventing the spread of infectious diseases.

Interaction of bio-objects and microbes with nano-particles can greatly affect their size, shape, chemical composition, surface and internal structure, charge, solubility and aggregation. For example, nano particles can be used to produce extraordinary images of tumor cells; single-walled carbon nano-tubes, act as highly efficient delivery agent for transport of bio-molecules into cells. Interlinking of nanotechnology with other fields like bio-based technologies, tele-communications and networking within human cells etc. to manipulate and transfer genetic material, and nano materials at smaller scale is very revolutionary in medical field like stem cell research, surgeries, monitoring tumors within body etc.

Immobilization of chemokine and tumor target agent on the surface of chitosan capped quantum dots with appropriate surface modification was reported by D. K. Chatterjee and Y. Zhang. Upon exposure to UV-light, quantum dots glow and give brighter image of cancer cells to visualize within the body for surgery. Thus nanotechnology is found to be very helpful for regeneration of injured nervous system.

Early diagnosis for Alzheimer's disease (AD) is essential for cure. As neuron degradation process started well before AD symptoms appears. Two nano-techniques bio-barcode assay (BCA) and localized surface Plasmon resonance (LSPR) were introduced as best for early diagnosis of AD after two articles were published in February 2005.

Silver nano-particles of 1–10 nm get attached to the surface of cell membrane and violently disturb bacterial functions like permeability and respiration etc. Due to small size, Ag nano-particle can penetrate easily inside the bacteria and rapidly interact with sulfur and phosphorus content based compounds like DNA for complete lyses of the bacterial cells. Silver nano-particles exhibit strong anti-bacterial properties by modulating the cellular signals due to dephosphotyrosine profile of bacterial peptides and by supporting and penetrating through bacterial cell wall to inhibit the microbial growth. This dose dependent effect is more specified against gram negative bacteria than gram-positive. The antibacterial effect of nano-particles is independent of resistance offered by the bacteria against antibiotics (Brunetti et al. 2007).

Carbon nano-tubes (CNTs) exhibited unique properties and wide range of applications in nano-medicine. Single-walled nano-tubes (SWNTs) is simply a rolling of simply a rolled-up grapheme sheet into a cylinder by semi-fullerene caps having the diameter lies within the range 0.8–2 nm. MWNTs are multilayered rolled up assays having the diameters of more than 50 to sometimes even 100 nm. Research implies that besides the huge variety of applications and interesting optical, mechanical and electrical properties, it is difficult to handle CNTs, otherwise may cause serious issues technically. The ability of solubilization, separation of single CNTs from their tightly bound bundles and purification make them more promising materials for many applications at nano-scale.

Modification of CNT's by introducing covalent interlinking at side-walls and tip is one of the robust techniques to improve the handling issues. For example, i) esterification of oxidized nano-tubes and (ii) direct attachment of functional groups by cyclic, nucleophilic, electrophilic or radical additions. CNTs can be synthesized and purified by arc discharge, laser ablation and catalyst based gas-phase growth by using carbon monoxide or other carbon sources as precursors. The contaminated CNTs have various amounts of amorphous carbon and/or catalytic metal particles. Such impure CNT's materials are treated under strong acidic and oxidative conditions. Specifically, the sonication has been done in a mixture of highly concentrated nitric and sulfuric acid or heating in a mixture of sulfuric acid and hydrogen peroxide. This process purifies the CNT's into shortened and open tubes. Carboxyl groups are produced at the caps and side walls as a result of this acidic oxidation process. Since the acid groups can react readily with alcohols or amines to give rise to ester or amide bonds, respectively. Functionalized and modified CNTs with organic and inorganic groups are very attractive tools for drug delivery, therapies, sensing and composite formation for vast applications (Zharov et al. 2003).

In recent years, the rapid development in surface microscopy, bio-physics and chemistry, computational networking together with nano-technology provides remarkable medical applications for twenty-first century.

Nanotechnology can create smart materials and drugs that may cure the people faster without any side effects. Energy production and storage is more effective in nano-drug systems. Nano techniques are more fast, portable and easy to handle, when compared with traditional drug injection methods. Parentally the nanotechnology is expensive one and creating more powerful and destructive atomic weapons.

5.5 Application of Nanotechnology in Drug Delivery

For applications to medicine and physiology, these nano materials, nanoparticles and devices can be designed to interact with cells and tissues at a molecular (i.e., subcellular) level with a high degree of functional specificity that allows the technology and biological systems integration. Unlike of traditional sciences like chemistry, biology, physics, material sciences, nanotechnology is multidisciplinary approach, requires collective aptitude for new developments.

Different types of materials and molecular level assemblies at nano-scale exhibit unique and advance physical and chemical properties for developing diagnostic and therapeutic treatments. Smart nanostructures, functionalized surfaces, scaffolds play powerful role in nano-medicine especially the drug delivery system etc. Nanoparticles occur in variety of shapes, texture, morphologies and specifically the sizes to provide many potential applications. Nanoparticles within size range of 50–100 nm are very suitable to move into the tumor call to cure the cancer. Nano-structures of 100–200 nm have low solubility, found to be more effective for drugs capping. Long chain polymers are strong enough, extended to high degree of precision are used to fabricate nano-bio efficient devices. Ligands on nanoparticle surfaces are used as target receptors to diagnose and cure the destructive tissue by attaching to it and releasing the drugs on targeted area. Bucky ball-based nano capsules may provide the treatment for AIDS in future. Cardio vascular disease can be cured by smart bio-active nano-materials by immobilizing the stem cells. Semi-invasive drug delivery devices produce the guiding images of implantations neuro-stimulators for delivery of neuro active molecules to the specific sites to cure neuro-degenerative diseases. Functionalization of 2-D and 3-D materials for spatial release of bio-chemical factors for artificial pancreas to measure and deliver the insulin by nano-enabled devices for controlling the diabetes.

Carbon nanotubes and *inorganic nanowires* display remarkable mechanical and magnetic strength, electrical conductivity and optical interactions that only few materials at this nano-platform could ever hope to match. CNT's and magnetic iron oxide nanoparticles, gold-coated silica nano shells can convert the electro-magnetic energy into heat increases the temperature to cancer cells by increasing the magnetic

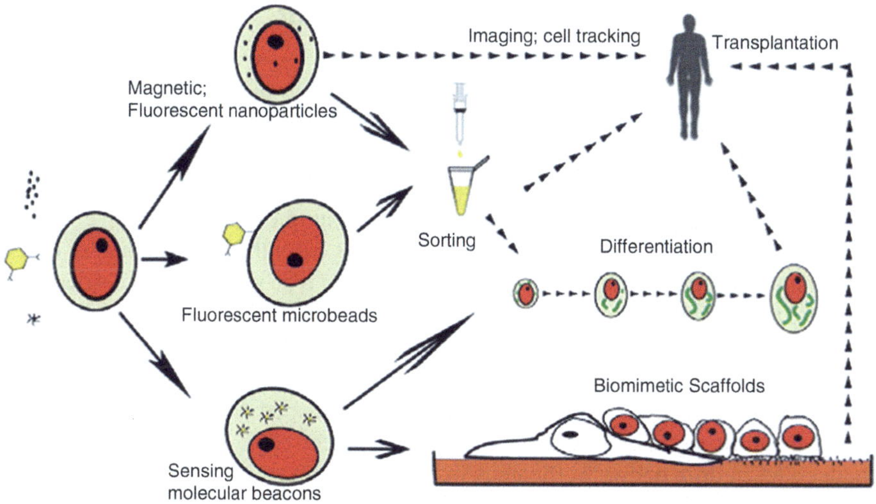

Fig. 5.2 Schematic representation of stem cells modulated with nano-carriers for tracking and imaging in transplantation within the human cells (Deb et al. 2012)

field or by laser irradiations at binding site of nanoparticles and tumor cells (Zharov et al. 2003).

Quantum dots are nano-sized semiconductor crystals with extraordinary fluorescence. Quantum dots exhibit tunable optical and electronic properties by changing their size and composition, due to their very small size (2–10 nm). Quantum dots are proof to be very demanding already in bio-nano imaging applications due to their small size within the range of 2–10 nm, low-cost and simple synthesis techniques. Quantum dots have ability to emit the light at any wavelength from visible and to infrared range on the spectrum, and can be inserted almost anywhere, including liquid solution, dyes etc. Quantum dots can be attached to a variety of surface ligands, and inserted into a variety of organisms for in-vivo research (Edina et al. 2014). Nano medicine provide early detection and prevention, improved diagnosis, proper treatment and follow-up of diseases become easy and approachable. Some nano-particles are also used as tag or label to perform biological process efficiently to make testing more sensitive and easy. Gold nanoparticles can be attached and tagged with segmented DNA for gene sequence detection within the living sample. With the help of artificially stimulated cells that might encourage the organ transplantation (Fig. 5.2).

Surface-lipid nano-particles SLN's have numerous applications, especially in cancer treatment: 1: chemotherapeutic drugs incorporated into SLNs to eliminate their disadvantages and overcome MDR, 2: gene therapy (delivery of peptides, genes, miRNA, siRNA into cells), 3: drug targeting, new adjuvants for vaccines, treatment of infectious diseases, cosmetic and dermatological preparations and agriculture applications (Darbre and Reymond 2006).

5.5.1 Nanotechnology for Cancer Treatment

Nanoparticle and ethylene glycol molecules attached to damaged cell and deliver the drugs to cancer tumors. Ethylene glycol prevents the white blood cells to identify the nanoparticles as foreign materials. Nanoparticles are allowed travel in the blood stream long enough to attach to cancer tumors. Research is going on to target the nano-encapsulated drugs directly to malfunctioning cells. For example, scientists have introduced that two types of nanoparticles carry the large amount of drug. The first type locates the site of cancer and the second type serve as drug carrier. The use of DNA segments to construct pores is in testing mode to carry the drug molecules.

Disk and rod-shaped nano-particles are found to be more active in drug- delivery as stick to the tumor surface longer than a spherical shaped nano-particle. Whereas, rod shaped are specifically effective for breast cancer cells. Thus, size and shape matter a lot in drug delivery efficiency in bio-medicine.

Gold nano-particles are attached to platinum containing molecule that has low toxicity for delivering the platinum to respective damage area. Platinum bearing nano-particle encounters an acidic solution in cancer affected area which changes the platinum to its toxic state for killing the cancer cells (Boysen 2016). Gold nano-particles are used for the treatment of skin damaging cancer cells, as on penetration, the RNA attaches to the cancer cell, thus, stop the generation of protein that cause the cancel cell growth. Nano-particles may also be used to deliver nitric oxide directly to cancer cells that reduce the amount of chemotherapy drugs. Elastic embedded carbon nano-fibers can penetrate into diseased cells to deliver the therapeutic drugs. Testing of nano-particles to deliver three anti-cancer drugs to the lymph nodes is also in process, but in animal lab yet. Significant reduction in tumor size was reported by injected hydrogel under the skin, as tested in laboratory mice for continuous drug release instead of repeated injection (Tiwari et al. 2012).

5.5.2 Nanotechnology for Heart Disease

A protein is attached to spoiled arteries for direct applications of drugs. The clot cracking drug was attached to nanoparticles clusters that break apart in regions of turbulent blood flow, in same was as blood flow is prevented by a clot. Polymer nanoparticles are placed on inflamed tissue such as arterial panel and dissolve and release the dug in the presence of hydrogen peroxide. Drugs can be delivered directly to stents placed in arteries by directing the iron-oxide nanoparticles by magnetic field.

5.5.3 Nanotechnology for Aging and Micillineous

Nano-diamonds are embedded in contact lenses for the treatment of glaucoma. The drug molecules are released from the nano-diamonds in contact with tears, provides continuous dosing than often occurs using eye drops. Addition of nanotubes to the surface of the implant material may improve the dental implantation. Blood sugar level can be well controlled for several days by using nano-particles that release insulin. Sponge like matrix that contains insulin and nano encapsulated enzymes is another way to release insulin. Rise in glucose concentration signaling the nano-capsules to release hydrogen ions that bind to the fibers making up the matrix. Hydrogen ions make the fibers positively charged and create the openings in the matrix to release the insulin outward. Nanoparticles frequently deliver antigens for a particular disease into the blood stream to reset the immune system by ceasing the white blood cells to destroy healthy cells. Nanoparticles can also deliver the drugs across the brain barrier to tackle neurologic disorders. Mesoporous nano-particles with various coatings releases the contents of nanoparticles to an enzyme found in aging cells, thus used as aging fighters. Such types of nanoparticles contents are added into creams and ointment etc. liposomes encapsulated in liposomes that merged with skin cells membranes and allow the delivery of proteins. Treatment of lung tissues is possible now by using nano-particles (Boysen 2016).

5.6 Future of Nano-drugs

Nanotechnology is already proven to be remarkable for therapeutics and diagnostics in medical science. Nano-scale drug delivery system will provide new innovations to meet up medical needs in cancer and other uncured disease. Further, new nano-imaging agents and smart materials will be pre-requires for execution of personalized medicine.

European technology platform on nano medicine (ETPN) indicates many challenges to be discussed: 1: fruitful interactions between academic and pharmaceutical industries, clinical organizations and medical agencies are necessary to develop the nano medicine. 2: Interlinking of academic innovations at smaller scale to drive large commercial scale in industries by sustainable R&D chain. Nanotechnology has outstanding impact on many different medical developments in the three main areas: therapeutics, diagnostics and imaging and regenerative medicine.

Tissue engineering will surely benefit the future nano medical technologies such as self-assembled and self-repaired soft nano materials composites with drugs and biological species have the potential to overcome the lack of migration of cells into template networking. The upcoming concept of matrix polymers specifically the hydro-gels enables the in situ cross-linking in the presence of effective cells. Nanotechnology is focusing on the surgically connectable vascularized matrices requires for implantation of tissue-engineered build ups on large scale (Tiwari et al.

2012). Prior diagnoses of diseases will be revolutionizing by multi-functional nano-particles that may perform both imaging and drug carrier features. Cancer is a big cause of death needs new therapeutic approaches that can directly target the disease without producing the side effects (Neal et al. 2004). Highly efficient and precise analysis, early detection and diagnosis, clear monitoring from nano-devices will definitely improve the patient's quality of life and lowers the mortality rates, in diseases like cancer and Alzheimer's disease.

5.7 Nano Drugs and Safety

To date, various nano drug systems are developed for different routes of administration include dendrimers, nano crystals, emulsions, liposomes, solid lipid nano-particles, micelles, and polymeric nano-particles as described above in detail.

Drug delivery systems based on nano-materials are retained to improve the efficacy, safety, physicochemical properties, and pharmacodynamics profile of pharmaceutical substances. Modified nano drug systems increases the bioavailability of orally taken drugs, long-ranged half-life of injected drugs (by reducing immunogenicity), and direct targeted delivery to defined tissues. Nano-materials are expected to lessen the toxicity of chemotherapy drugs and other drugs with narrow therapeutic index. Despite of all pharmacokinetic advantages, nano drugs may create some toxic potential such as in vitro and in vivo cyto-toxicity, oxidative stress due to release of oxygen species and free radicals, inflammation, DNA damaging, formation of multi-nuclei and fibrosis etc. Safe and impressive development in nano drug system requires better understanding of the pharmacokinetic and limitations of each delivery option.

In pharmaceutical research and development, significant drug carriers must have (1): high solubility; (2) efficient membrane permeability; (3) metabolic stability; and (4) systemic pharmacokinetics. Various nano drug techniques might resolve the biopharmaceutical issues by controlling the quantity of drug release, poor stability, limited pharmacokinetic behavior, and toxicity of the active ingredient. Recently as much as ~70% of new drug candidates have shown poor solubility in water media, and ~40% of marketed drugs in oral use are identified to be practically insoluble in aqueous media (100 μg/mL) (Herrero et al. 2009).

The challenges posed by nano drugs in medicine will only increase over the next 20 years as combination therapies confuse the interface between drug and device. Although a cure for all cancers may always remain difficult to catch but it is still realistic to expect hallmarks from the nano pharmaceutical field.

References

Baranello MP, Bauer L, Benoit DS (2014) Poly(styrene-*alt*-maleic anhydride)-based diblock copolymer micelles exhibit versatile hydrophobic drug loading, drug-dependent release, and internalization by multidrug resistant ovarian cancer cells. Biomacromolecules 15(7):2629–2641

Boulaiz H, Alvarez PJ, Ramirez A, Marchal JA, Prados J, Rodríguez-Serrano F, Perán M, Melguizo C, Aranega A (2011) Nanomedicine: Application areas and development prospects. Int J Mol Sci 12(5):3303–3321

Boysen E of Hawk's Perch Technical Writing, LLC and UnderstandingNano.com. Copyright ©20072016 Hawk's Perch Technical Writing, LLC-All Rights Reserved

Bradbury J (2003) Nanoshell destruction of inoperable tumours. Lancet Oncol 4:711

Britten MB, Abolmaali ND, Assmus B, Lehmann R, Honold J, Schmitt J, Vogl TJ, Martin H, Schachinger V, Dimmeler S, Zeiher AM (2003) Infarct remodeling after intracoronary progenitor cell treatment in patients with acute myocardial infarction (TOPCARE-AMI) mechanistic insights from serial contrast-enhanced magnetic resonance imaging. Circulation 108:2212–2218

Brunetti FG, Herrero MA, Muñoz JM, Giordani S, Díaz-Ortiz A, Filippone S, Ruaro G, Meneghetti M, Prato M, Vázquez E (2007) Reversible microwave-assisted cycloaddition of aziridines to carbon nanotubes. J Am Chem Soc 129:14580–14581

Carlmark A, Hawker C, Hult A, Malkoch M (2009) New methodologies in the construction of dendritic materials. Chem Soc Rev 38:352–362

Cho H, Kwon GS (2014) Thermosensitive poly-(d,l-lactide-co-glycolide)-block-poly(ethylene glycol)-block-poly-(d,l-lactide-co-glycolide) hydrogels for multi-drug delivery. J Drug Target 22:669–677

Cho H, Lai TC, Kwon GS (2013) Poly(ethylene glycol)-*block*-poly(ε-caprolactone) micelles for combination drug delivery: Evaluation of paclitaxel, cyclopamine and gossypol in intraperitoneal xenograft models of ovarian cancer. J Contr Release 166(**1**):1–9

Cho H, Lai TC, Tomoda K, Kwon GS (2014) Polymeric micelles for multi-drug delivery in cancer. AAPS Pharm Sci Tech 16:10–20

Darbre T, Reymond J (2006) Peptide dendrimers as artificial enzymes, receptors, and drug-delivery agents. Acc Chem Res 39(12):925–934

Deb KD, Griffith M, Muinck ED, Rafat M (2012) Nanotechnology in stem cells research: advances and applications. Front Biosci (Landmark Ed) 17:1747–1760

Dikmen G, Genç L, Güney G (2011) Advantage and disadvantage in drug delivery system. J Mater Sci Eng 5(4):468–472

Dufes C, Uchegbu IF, Schatzlein AG (2005) Dendrimers in gene delivery. Adv Drug Delivery Rev 57:2177–2202

Edina CW, Andrew ZW (2014) Nanoparticles and their applications in cell and molecular biology. Integr Biol (Camb) 6(1):9–26

Emeje MO, Obidike IC, Akpabio EI, Ofoefule SI (2012) Nanotechnology in drug delivery. In: Sezer AD (ed) InTech, Rijeka, Croatia, pp 70–106

Eskiler G G, Gökhan D, Lütfi G (2013) Nano-based drug delivery system, Intech. (2013):90–146. http://iapc-obp.com/assets/files/106072_08_NBDD_03.pdf

Estanqueiro M, Amaral MH, Conceiçao J, Lobo JMS (2015) Nanotechnological carriers for cancer chemotherapy: the state of the art. Colloids Surf B 126:631–648

Fukushima S, Varela-Carver A, Coppen S R, Yamahara K, Felkin L E, Lee J, Barton P J, Terracciano C M, Yacoub M H, Suzuki K (2007) Direct intramyocardial but not intracoronary injection of bone marrow cells induces ventricular arrhythmias in a rat chronic ischemic heart failure model. Circulation 115(17):2254–2261

Gaucher G, Satturwar P, Jones MC, Furtos A, Leroux JC (2010) Polymeric micelles for oral drug delivery. Eur J Pharm Biopharm 76:147–158

Güney G, Genç L, Dikmen G (2011) J Mater Sci Eng 5(5):577–582

Gupta A, Arora A, Menakshi A, Sehgal A, Sehgal R (2012) nanotechnology and its applications in drug delivery: a review. Int J Med. Mol Med 3(1):1–9

Haley B, Frenkel E (2008) Nanoparticles for drug delivery in cancer treatment. Urol Oncol Semın Ori 26:57–64

Han Y, He Z, Schulz A, Bronich TK, Jordan R, Luxenhofer R, Kabanov AV (2012) Long-circulating PEG-PE micelles co-loaded with paclitaxel and elacridar (GG918) overcome multidrug resistance. Mol Pharm 9:2302–2313

Herrero MA, Toma FM, Al-Jamal KT, Kostarelos K, Bianco A, Ros TD, Bano F, Casalis L, Scoles G, Prato M (2009) Synthesis and characterization of a carbon nanotube–dendron series for efficient siRNA delivery. J Am Chem Soc 131(28):9843–9848

Hornyak GL, Tibbals HF, Dutta J, Moore JJ (2008) Introduction of nano science and nanotechnology. Chapter no: 9, Carob-based nanomaterials, December 22. CRC Press, Boca Raton

http://www.understandingnano.com/cancer-therapy-platinum-gold-nanoparticles.html

http://www.understandingnano.com/nanotechnology-drug-delivery.html

Jaafar-Maalej C, Diab R, Andrieu V, Elaissari A, Fessi H (2010) Ethanol injection method for hydrophilic and lipophilic drug-loaded liposome preparation. J Liposome Res 20(3):228–243

Katayose S, Kataoka K (1998) Remarkable increase in nuclease resistance of plasmid DNA through supramolecular assembly with poly (ethylene glycol)—poly (L-lysine) block copolymer. J Pharm Sci 87:160–163

Kesharwani P, Jain K, Jain NK (2014) Dendrimer as nanocarrier for drug delivery. Prog Polym Sci 39(2):268–307

Kozlov M, Melik-Nubarov N, Batrakova E, Kabanov A (2011) Relationship between pluronic block copolymer structure, critical micellization concentration and partitioning coefficients of low molecular mass solutes. Macromolecules 33:3305–3313

Krishnamurthy S, Ng VW, Gao S, Tan MH, Yang YY (2014) Phenformin-loaded polymeric micelles for targeting both cancer cells and cancer stem cells *in vitro* and *in vivo*. Biomaterials 35(33):9177–9186

Kumar R, Kulkarni A, Nagesha DK, Sridhar S (2012) *In vitro* evaluation of theranostic polymeric micelles for imaging and drug delivery in cancer. Theranostics 2(7):714–722

Kutty RV, Feng SS (2013) Cetuximab conjugated vitamin E TPGS micelles for targeted delivery of docetaxel for treatment of triple negative breast cancers. Biomaterials 34(38):10160–10171

Laouini A, Jaafar-Maalej C, Limayem-Blouza I, Sfar S, Charcosset C, Fessi H (2012) Preparation, Characterization and Applications of Liposomes: State of the Art. J Colloids Sci Biotechnol 1:147–168

Lee H, Soo PL, Liu J, Butler M, Allen C (2007) Polymeric micelles for formulation of anti-cancer drugs. In: Amiji MM (ed) CRC Press, New York, pp 318–355

Lencioni R, Bartolozzi C (1996) Ethanol injection for the treatment of hepatic tumours. Eur Radiol 6(5):682–696

Lopes S C A, Giuberti C S, Rocha T G R, Ferreira D S, Leite E A, Oliveira M C (2013). Liposomes as carrier of anticancer drugs, L. Rangel (Ed.), Intech.

Mader K, Mehnert W (2004). Solid lipid nanoparticles – concepts, procedures and physicochemical aspects, C. Nastruzzi (Ed.), CRC Press, Boca Rotan: 1–22.

Makino K, Shibata A (2006). Surface properties of liposomes depending on their composition. In: Liu AL (ed) Elsevier, New York, pp 49–53

Malik A, Chaudhary S, Garg G, Tomar A (2012) Dendrimers: a tool for drug delivery. Adv Biol Res 6(4):165–169

Molina EJ, Palma J, Gupta D, Torres D, Gaughan JP, Houser S, Macha S, Thorac J (2008) Improvement in hemodynamic performance, exercise capacity, inflammatory profile, and left ventricular reverse remodeling after intracoronary delivery of mesenchymal stem cells in an experimental model of pressure overload hypertrophy. Cardiovasc Surg 135:292–299

Molina EJ, Palma J, Gupta D, Gaughan JP, Houser S, Macha M (2009) Right ventricular effects of intracoronary delivery of mesenchymal stem cells (MSC) in an animal model of pressure overload heart failure. Biomed Pharmacother 63:767–772

Mussi SV, Torchilin VP (2013) Recent trends in the use of lipidic nanoparticles as pharmaceutical carriers for cancer therapy and diagnostics. J Mater Chem B 1:5201–5209
Na HS, Lim YK, Jeong YI, Lee HS, Lim YJ, Kang MS, Cho CS, Lee HC (2010) Combination antitumor effects of micelle-loaded anticancer drugs in a CT-26 murine colorectal carcinoma model. Int J Pharm 383(1–2):192–200
Nakamura K, Ryuichi E, Takeda M (1976) Surface properties of styrene – ethylene oxide block copolymers. J Polym Sci 14(7):1287–1295
Nasibullah M, Hassan F, Ahmad N, Khan AR, Rahman M (2013) Dendrimers as novel polymeric material: a review on its synthesis, characterization and their applications. Adv Sci Focus 1(3):197–204
Neal DP, Hirsch LR, Halas NJ, Payne JD, West JL (2004) Photo-thermal tumor ablation in mice using near infrared-absorbing nanoparticles. Cancer Lett 209(2):171–176
Patil J, Gurav P, Kulkarni R, Jadhav S, Mandave S, Shete S, Chipade V (2013) Applications of solid lipid nanoparticle in novel drug delivery system. Brit. Biomed Bull 1(2):103–118
Ramandeep Singh MK (2016) Kale and Atul Bodkhe, Liposomes: from concept to commercialization. Euorpian J Biomed Pharm sci 3(7):189–206
Ramasamy T, Kim J, Choi HG, Yong CS, Kim JOJ (2014) Novel dual drug-loaded block monomer complex micelles for enhancing the efficacy of chemotherapy treatments. Nanotechnology 10(7):1304–1312
Sajomsang W, Gonil P, Saesoo S, Ruktanonchai UR, Srinuanchai W, Puttipipatkhachorn S (2014) Synthesis and anticervical cancer activity of novel pH responsive micelles for oral curcumin delivery. Int J Pharm 477(1–2):261–272
Sanganalmath SK, Bolli R (2013) Cell therapy for heart failure: a comprehensive overview of experimental and clinical studies, current challenges, and future directions. Circ Res 113:810–834
Sarisozen C, Vural I, Levchenko T, Hincal AA, Torchilin VP (2012) Long-circulating PEG-PE micelles co-loaded with paclitaxel and elacridar (GG918) overcome multidrug resistance. Drug Deliv 19(8):363–370
Satchi-Fainaro R, Duncan R, BarnesB C M (2006) Polymer therapeutics for cancer: current status and future challenges. In: Satchi-Fainaro R, Duncan R (eds) Springer, New York, pp 1–65
Schwarz MA, Raith K, Neubert RHH (1998) Characterization of micelles by capillary electrophoresis. Electrophoresis 19:2145–2150
Shen Y, Pei A, Sun A, Xu J, Song Y, Huang G, Sun X, Zhang S, Qin Q, Zhu H, Yang S, Yang X, Zou Y, Qian J, Ge J (2013) Comparison of magnetic intensities for mesenchymal stem cell targeting therapy on ischemic myocardial repair: high magnetic intensity improves cell retention but has no additional functional benefit. Biomaterials 34:9905–9916
Shi S, Shi K, Tan L, Qu Y, Shen G, Chu B, Zhang S, Su X, Li X, Wei Y, Qian Z (2014) The use of cationic MPEG-PCL-g-PEI micelles for co-delivery of Msurvivin T34A gene and doxorubicin. Biomaterials 35(15):4536–4547
Singh SK, Sharma VK (2013) Dendrimers: A class of polymer in the nanotechnology for drug delivery. In: AK Mishra (ed), Scrivener Publishing, Beverly
Sosnik A, Raskin MM (2015) Polymeric micelles in mucosal drug delivery: Challenges towards clinical translation. Biotechnol Adv 33(6):1380–1392
Sun CY, Ma YC, Cao ZY, Li DD, Fan F, Wang JX, Tao W, Yang XZ (2014) Effect of Hydrophobicity of Core on the Anticancer Efficiency of Micelles as Drug Delivery Carriers. ACS Appl Mater Interfaces 6(24):22709–22718
Tiwari G, Tiwari R, Sriwastawa B, Bhati L, Pandey S, Pandey P, Saurabh KB (2012) Drug delivery systems: an updated review. Int J Pharm Investig 2(1):2–11
Tripathy S, Das MK (2013) Dendrimers and their applications as novel drug delivery carriers. J Appl Pharmaceut Sci 3(9):142–149
Vemuri S, Rhodes CT (1995) Preparation and characterization of liposomes as therapeutic delivery systems: a review. Pharm Acta Helv 70(2):95–111
Venkataraman S, Hedrick SL, Ong ZY, Yang C, Pui Lai Rachel Ee, Hammond PT, Yang YY (2011) The effects of polymeric nanostructure shape on drug delivery. Adv Drug Deliv Rev 63(14-15):1228–1246

Wang H, Zhao Y, Wu Y, Hu YL, Nan K, Nie G (2011) Enhanced anti-tumor efficacy by co-delivery of doxorubicin and paclitaxel with amphiphilic methoxy PEG-PLGA copolymer nanoparticles. Biomaterials 32:8281–8290

Weissig V, Whiteman KR, Torchilin VP (1998) Accumulation of protein-loaded long-circulating micelles and liposomes in subcutaneous lewis lung carcinoma in mice. Pharm Res 15:1552–1556

Welch PM, Welch CF (2009) Tecto-dendrimers: a study of covalently bound nanospheres. Macromolecules 42:7571–7578

Wilczewska AZ, Niemirowicz K, Markiewicz KH, Car H (2012) Nanoparticles as drug delivary systems. Pharmacol Rep 64:1020–1037

Xie X, Sun A, Zhu W, Huang Z, Hu X, Jia J, Zou Y, Ge J (2012) Transplantation of mesenchymal stem cells preconditioned with hydrogen sulfide enhances repair of myocardial infarction in rats. Tohoku J Exp Med 226:29–36

Yhee J Y, Son S, Son S, Joo MK, Kwon IC (2013). The EPR effect in cancer therapy. In: Bae YH, Mrsny RJ, Park K (eds) Springer, New York, pp 621–632

Yokoyama M (2010) Polymeric micelles as a new drug carrier system and their required considerations for clinical trials. Expert Opin Drug Deliv 7:145–158

Zhang X, Liu B, Yang Z, Li H, Luo X, Luo H, Gao D, Jiang Q, Liu J, Jiang Z (2014a) Micelles of enzymatically synthesized PEG-poly(amine-*co*-ester) block copolymers as pH-responsive nanocarriers for docetaxel delivery. Colloids Surf B 115:349–358

Zhang X, Zeng X, Liang X, Yang Y, Li X, Chen H, Huang L, Mei L, Feng S S (2014b) The chemotherapeutic potential of PEG-*b*-PLGA copolymer micelles that combine chloroquine as autophagy inhibitor and docetaxel as an anti-cancer drug. Biomaterials 35(33):9144–9154

Zharov V, Galitovsky V, Viegas M (2003) Photothermal detection of local thermal effects during selective nanophotothermolysis. Appl Phys Lett 83(24):4897–4899

Chapter 6
Protein Capping and Nanoparticles

Nadia Ghaffar and Shagufta Naz

6.1 General Introduction

Nanoscience is accepted as an accomplished field of science that covers various scientific fields, like biology, physics and chemistry including materials science. Nanotechnology is one of the most exciting fields of nanoscience which plays an important role in solving many complex technical problems of current science, primarily in biomedical sciences and nanomedicine. Nanoparticles are considered vital in medical biology because of their exceptional interaction with cells, due to their small size and potential to target even brain (Mahmoudi et al. 2011a, b, c, d; Ragnaill et al. 2011). Therefore, NPs show many applications such as gene delivery (Tanaka et al. 2010; Ditto et al. 2009), drug delivery, tissue repair, disease diagnosis, cancer therapy (Liang et al. 2010), magnetic resonance spectroscopy (Kircher et al. 2012) and hyperthermia (Lee et al. 2011). NPs contain extremely active surface chemistry due to their large surface-to-volume ratio. Due to small size of NPs, they may come in contact with any biomolecules such as lipids, proteins, nucleic acids and biological metabolites. When nanoparticles interact with any biological environment, they adsorb biomolecules by lowering their high surface energy, as a result intricate layer of biomolecules is formed that covers the nanoparticle's surface.

One of the examples of such biomolecules is the 'proteins'. More than 10,000 kinds of protein are found in mammalian serum/plasma. These proteins interact with nanoparticles when they are in contact. Most important is the adsorption of a particular protein on the surface of nanoparticles, which forms nanoparticle-protein complexes that is usually considered as nanoparticle-protein corona (NP-PC).

N. Ghaffar
Department of Botany, Lahore College for Women University, Lahore, Pakistan

S. Naz (✉)
Department of Biotechnology, Lahore College for Women University, Lahore, Pakistan

S. Javad, A. Butt (eds.), *Nanobotany*, https://doi.org/10.1007/978-3-319-77119-9_6

Generally this protein capping is responsible for biological reactivity of nanoparticles (Casals et al. 2010; Cedervall et al. 2007a). The protein capping depends on different kinds of protein that compete with each other to adsorb on surface of NPs. Miclaus et al. proposed that the nanoparticle-corona complex is dynamic and after its interaction with cellular system, it is the first and new entity "what the cell sees" (Miclaus et al. 2014; Walczyk et al. 2010).

It is considered that interaction of protein with NPs having planar surfaces causes the distortion of protein after adsorption on the surface of nanoparticles. Protein capping on nanoparticles is complex process and no plasma protein is considered universal for corona formation for all NPs and densities of proteins are also independent to their abundance in plasma. So, protein capping is unique for each nanoparticle.

6.2 Nanoparticles Protein Capping Structure and Composition

After penetration of nanoparticles in a biological fluid, different biomolecules like proteins, polysaccharides, lipid and nucleic acids adsorb on the surface of NPs. Proteins are consist of polypeptides that carry a net surface charge with distinct conformation influenced by pH of the medium. Many forces such as hydrogen bonding, Vander Waals interactions and solvation forces support the binding of proteins during the interaction with NPs. Protein capping on nanoparticles is a complex process and depends upon three factors i.e., characteristics of the NP, medium of interaction and interacting proteins. It must be well understood that organization, composition and time of substitution of proteins on NPs is read by the cell. Usually proteins are present in high concentrations and also having good association first arrive and occupy the surface of NPs, but quickly dissociate and replaced by proteins with high-affinity and present in lower concentration (Hussain et al. 2005).

6.3 Types of Corona

Roughly corona is divided into two types, hard and soft corona.

6.3.1 Hard Corona

In hard corona proteins adsorb to the surface strongly and tightly. It is a dynamic process and with the passage of time exchange of proteins takes place in the media (Eckhardt et al. 2013). Generally higher affinity proteins exchange slowly and form

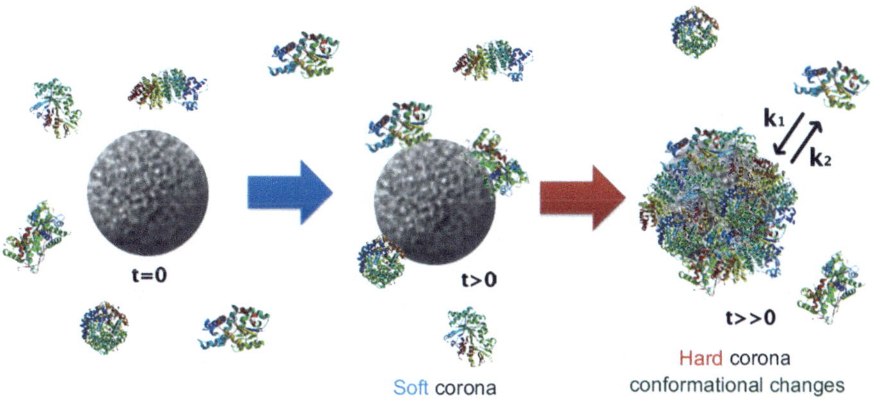

Corona	Soft	Hard
K_d (dissociation constant)[9,20,54]	High	Low
Adhesion to hydrophobicity[35]	Low	High
Molecular weight[25,32]	Low	High
Endosome–Lysosome trafficking[13,21]	Low*	Low*, High†
Conformational changes (sheet)[35,43]	Low	High

Fig. 6.1 Representation and characteristics of soft and hard corona. Nanoparticles are surrounded by capping proteins. Hard corona has high affinity for the surface of nanoparticles (For long period, even hours) and soft corona has less affinity for the surface of NPs (For a short period, even seconds or minutes). *Compared with serum-free condition. †Compared with soft corona. (Modified from Ref. Lee et al. 2011)

hard corona. Usually the inner layer consists of tightly bound proteins. Hard corona have high affinity for the surface of nanoparticles (For long period, even hours) (Fig. 6.1). The binding of proteins with nanoparticles may cause a change in the innate properties of both proteins and nanoparticles (Miclaus et al. 2014).

6.3.2 *Soft Corona*

In soft corona proteins are bound to the surface less tightly. Basically low affinity proteins are replaced quickly and form soft corona. This shows that outer layer consist of loosely bound proteins. Soft corona has less affinity for the surface of NPs (For a short period, even seconds or minutes) (Fig. 6.1).

A review of literature shows that in the binding of proteins with nanoparticles, most of the nanoparticles bind apolipoproteins (Cedervall et al. 2007b). In human plasma, proteins like immunoglobulins, fibrinogen, serum albumin, apolipoproteins etc. are present and involved in nanoparticles protein corona formation (Table 6.1).

Table 6.1 Detailed overview of serum/plasma proteins adsorbed on the surface of different types of nanomaterials with various sizes and surface chemistries

Nanomaterial used	Size (nm)	Surface chemistry	Dispersal medium	Proteins identified	References
Polystyrene NPs	50,100	NH_2, COOH	Human plasma	Coagulation factors, lipoproteins, acute phase proteins, immunoglobulins, complement proteins,	Lundqvist et al. (2008)
Polystyrene NPs	100	COOH	Human serum (depleted)	Complement proteins, immunoglobulin, albumin, complement, plasminogen, anti-CD4, c4a, plasminogen	Ehrenberg et al. (2009)
Latex NPs	80–109	NH_2,NHR, NR_2,NR^{3+} coo^-, so^{-3}, so^{-4}	Human serum	Albumin, hemoglobin, Haptoglobins, Apolipoproteins, immunoglobulins	Gessner et al. (2003)
Copolymer NPs	70, 200	–	Human plasma	Albumin, immunoglobulins, C4BPαchain, Apolipoprotiens, fibrinogen	Cedervall et al. (2007b)
MWCNTs	20–30	NH_2, COOH	Human plasma	2 macroglobulin precursor, coagulation factors, complement factors, plasminogen	Gasser et al. (2010)
SPIONs	–	–	Human plasma	Albumin, α1Antitrypsin, transferrin, Transthyretin, fibrinogen chains, immunoglobulin chains,	Jansch et al. (2012)
Gold	5, 10, 20	(PAA) polymer coated	Human plasma	Albumin, Apolipoprotein A1, fibrinogen chains	Deng et al. (2013)
Gold	15, 40, 80	–	Bovine serum	Transport proteins, tissue development proteins, coagulation factors	Maiorano et al. (2010)
TiO_2 NPs ZnO NPs quartz sand carbon nanotubes SiO_2 NPs	–	Silone, alumina, silica coated	Human plasma	Fibrinogen chains, albumin, ApoA1, immunoglobulin light chains, fibrin, complement component proteins, fibronectin	Sund et al. (2011)
SiO_2 NPs	8, 20, 25	–	Human plasma	Immuno globulins, coagulation proteins, lipoproteins, complement proteins, acute phase proteins, serum proteins, cellular proteins	Tenzer et al. (2011)
TiO_2 NPs ZnO NPs SiO_2 NPs	–	–	Human plasma	Albumin, fibrinogen, transferrin, Immuno globulins, complement proteins, Apolipoprotein A1	Deng et al. (2009)
Magnetic NPs	50, 200	Dextran COOH, NH_2, PEG COOH, PEG-NH_2	Bovine serum	Albumin, Vitronectin, Apolipoprotein A-1 complement factors, Haemoglobin	Mishra et al. (2009)

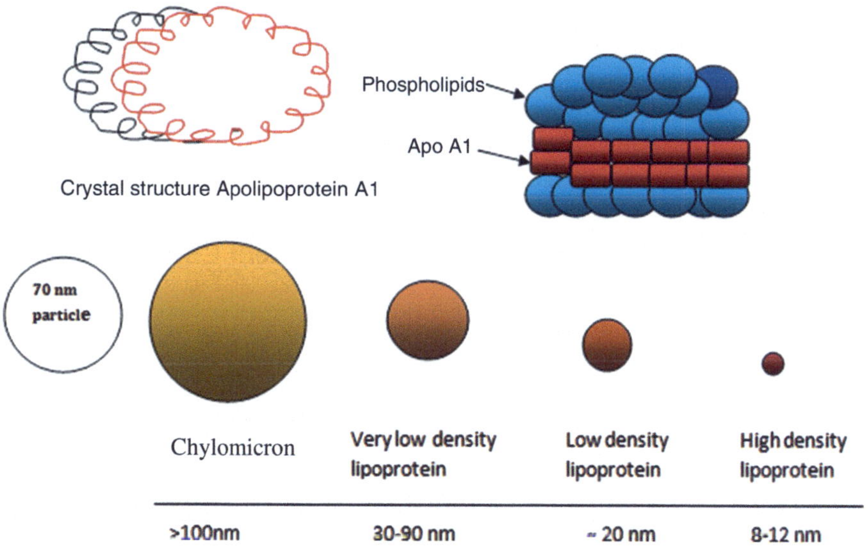

Fig. 6.2 (**a**) Crystal structure of the protein apolipoprotein A-1, and diagrammatic representation of lipoprotein complex containing phospholipids and apolipoprotein A-1. (**b**) Comparison of size of 70 nm nanoparticles with lipoprotein complexes (chylomicrons, very low, low, and high density lipoproteins)

In the starting it was a surprising result and distinct for the nanoparticles with flat surface. But, apolipoproteins are involved in the formation of lipoprotein complexes that contain size on nanoscale, from ~10 to ~100 nm (Fig. 6.2), which shows that interaction between nanoparticles and apolipoproteins is size dependent.

There are several factors that influence the capping of proteins on NPs. One of the factors that influence the capping is the amount of protein that is available during interaction of protein with the NP surface. Monopoli and coworkers suggested that protein adsorption to NPs changes with plasma concentration, binding of abundant proteins increased by increasing the concentration of plasma proteins (Monopoli et al. 2011).

In a survey, Walkey and Chan recapitulate a subset of 125 plasma proteins, named them **adsorbome**, which were known to be found in protein corona of NPs. Hellstrand and co-workers investigated that protein corona of polystyrene NPs contains high density lipoproteins (Hellstrand et al. 2009). The binding pattern of the proteins present in blood to NPs surfaces is dynamic, proteins present in high concentration like fibrinogen and albumin initially bound to NPs surface and then quickly replaced by high affinity proteins (Vroman 1962).

Simberg and co-workers worked on a model and introduced the term of "primary and secondary binders" for the protein corona, initially "primary binders" interact with the surface of NPs then "secondary binders" interact with primary through protein–protein interactions (Simberg et al. 2009). For the physiological response such a type of multilayered structure is very important because when secondary binders interacts, may cause a change in the functioning of primary binders or can "mask" them and effect their contact with the surrounding environment and may cease its all interaction. In majority of the plasma proteins a hydrodynamic diameter (3–15 nm) is observed; so, the protein capping on NPs is thick. The capping protein may contain single or multiple layers of binding proteins.

6.4 Nanoparticles Induced Changes in Protein Conformation

Proteins are consisting of chains of amino acids; protein's structure, function and shape is determined by the correct sequence of these amino acids. α-helices and β-sheets are the secondary structures of proteins and these two form a three-dimensional arrangement known as tertiary structure. The native structure of protein is strongly controlled by the shape of the hydrophobic residues that plays important role in close packing of the cores (Word et al. 1999).

Proteins undergo structural rearrangement when adsorb on the surface of nanoparticles, called conformational changes. In **irreversible conformational changes** the original structure of protein is not recovered after removal of protein from nanoparticles, while in **reversible conformational change**, protein retains its original structure. The nanoparticles surface plays an important role in modifying the structure and function of the binding protein consequently overall bio-reactivity of the NP may also changed. These changes may be positive as well as negative.

(1) It is considered that hydrophobic nanoparticles underwent more conformational change as compared to hydrophilic NPs.
(2) Curved surfaces NPs are extra flexible as compared to planar surfaces NPs and are significant in enhancing the surface area of the binding proteins (Verma and Stellacci 2010). Curved surfaces NPs play a significant role in irreversible changes and may also change the secondary structures of proteins (Worrall et al. 2006). Structural flexibility and chemical nature of proteins also play important role in their structural modification by nanoparticles (Lacerda et al. 2009).

It was investigated by Wangoo et al. that Gold NPs affect the conformational changes of bovine serum albumin (BSA) (Wangoo et al. 2008), while carbon C60 fullerene NP showed no conformational change after adsorption of proteins to NPs (Liu et al. 2012). Titanium dioxide (TiO_2) NPs were also investigated for a vital cytoskeletal protein and observed to influence the conformational changes and also cause reduction in polymerization of tubulin so that decrease the cell division and negative effect of cancer cells (Gheshlaghi et al. 2008). Interaction of zinc oxide NPs with BSA was also investigated and observed with no significant structural changes but minor conformational perturbation were examined (Bardhan et al. 2009). Transferrin proteins interact with SPIONs and undergo an irreversible conformational change in its secondary structure (Mahmoudi et al. 2011a, b, c, d). Protein conformational changes induced by NPs may influence the cellular signaling, protein protein interactions and DNA transcription, which is mostly vital for

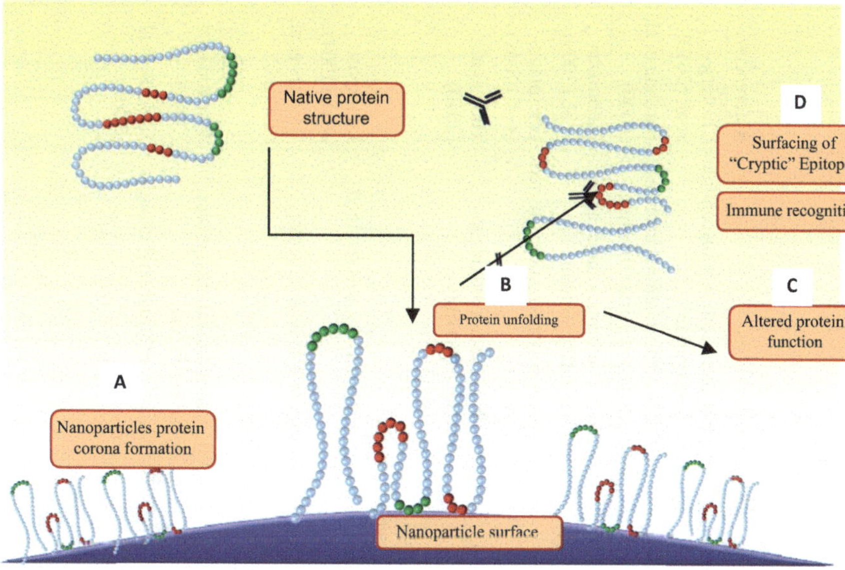

Fig. 6.3 Schematic illustration of protein unfolding after adsorption to surface of NP and consequences. (**A**) Adsorption of protein on NP surface (**B**) Formation of NP-PC. Surface of NP may influence abnormal unfolding of adsorbs proteins (**C**) cause unfolding of native structure of protein and changes its function (**D**) depicts hidden ("cryptic") epitopes

enzymes. NPs induced conformational changes in the active site may also cause loss of enzyme activity (Karajanagi et al. 2004).

Turci et al. observed that silica NPs showed no effect on RNAse and lysozymes while albumin and lactoperoxidase didn't retain their native structures and undergo an irreversible conformational change (Turci et al. 2010). It can be easily understood with the help of antibody and antigen reaction. Polypeptide chains of proteins undergo specific folding which results in conformational epitopes. Epitopes are the specific regions found on the primary structure of protein containing 10–12 amino acids and can accomplish immune response; basically epitopes are antigen binding sites that are recognized by antibodies. Adsorption of proteins on the surface of NPs causes formation of new conformational epitopes because of abnormal unfolding of adsorbs proteins and also depicts hidden epitopes by unfolding the native structure of protein. (Fig. 6.3). Function of adsorb proteins is greatly affected by these occult epitopes, e.g. eliciting unwanted enhanced or depressed immune response.

The surface of the nanoparticles can also induce thermodynamic instability in binding protein and thus causing chemical denaturation. ZnO NPs influence the unfolding of periplasmic domain of *Vibrio cholera* ToxR protein causing denaturation of protein by chaotropic agents (Chatterjee et al. 2010). Fascinatingly, ZnO NPs are also not in favor of denaturing agents and are capable of stabilizing the α-helical content of lysozyme (Chakraborti et al. 2009). To some extent proteins depend upon their chemical properties after adsorption to the surface of NPs (Table 6.2).

Table 6.2 Review of literature on NP induce conformational change to the adsorbed protein native structure

NP type and size	Protein investigated	Protein mol. wt. and size nm (if provided)	Change in protein structure	Analytical technique	Observations	References
ZnO NPs (25 nm)	*Vibrio cholera* Tox r	32.5 kDa	Yes	CD	NP-protein complex susceptible to denaturation	Chatterjee et al. (2010)
ZnO NPs (N/A)	BSA	66 kDa	Yes	CD	Minor conformational changes, secondary structure retained	Bardhan et al. (2009)
ZnO NPs (N/A)	BSA	66 kDa	Yes	FTIR	Minor conformational changes in secondary structure	Kathiravan et al. (2009)
TiO_2 NPs (20 nm)	Tubulin	55 kDa	Yes	FS	Protein polymerization affected	Gheshlaghi et al. (2008)
SiO_2 NPs (~40 nm)	BSA	66 kDa	Yes	RS	BSA and lactoperoxidase bound irreversibly	Turci et al. (2010)
	Hen egg lysozyme	14.3 kDa	No			
	RNASe A	13.7 kDa	No			
	Lactoperoxidase	77.5 kDa	Yes			
SiO_2 NPs (6,9,15 nm)	Human carbonic anhydrase	29 kDa	Yes	NMR	Protein activity was retained	Lundqvist et al. (2004)
Alumina and hydroxyapatite particles (100–300 nm)	BSA	66 kDa 8 × 8 × 3	Yes	FTIR	Loss in α-helical structure	Brandes et al. (2006)
	Hen egg lysozyme	14.3 kDa 4.6 × 3 × 3	Yes			
	Bovine serum fibrinogen	350 kDa 6 × 6.5 × 45	Yes			
Gold (45 nm)	BSA	66 kDa	Yes	CD	Conformational change was dose dependent	Wangoo et al. (2008)
Gold (5–100 nm)	Albumin	67 kDa	Yes	CD and FS	Minor conformational changes observed	Lacerda et al. (2009)
	Fibrinogen	340 kDa	Yes			
	ɣ-globulin	120 kDa	Yes			
	Histone H3	15 kDa	Yes			
	Insulin	5.8 kDa	Yes			

Gold (7–22 nm)	Human fibrinogen	340 kDa 45 × 5	Yes	CD	Unfolding induced immune response in THP-1 cells	Deng et al. (2011)
SPIONs (5–10 nm)	Transferrin	80 kDa	Yes	CD	Irreversible interaction	Mahmoudi et al. (2011a, b, c, d)
SWCNTs (N/A)	Horse radish peroxidase	44 kDa	No	CD	NP-protein complexes retained enzymatic activity	Asuri et al. (2007a, b)
	Subtilisin Carlsberg	39 kDa	No			
	Chicken egg white lysozyme	14.3 kDa	No			

Abbreviations used: *CD* Circular dichroism spectrometry, *FTIR* Fourier transformed infrared spectrometry, *FS* Fluorescence spectroscopy, *RS* Raman spectroscopy, *NMR* Nuclear magnetic resonance

6.5 Competitive Surface-Adsorption of Proteins with Time

When NPs enter in the body different biomolecules like proteins, polysaccharides, lipid and nucleic acid compete for the surface of NPs consequently soft and hard corona is formed with typical short and long exchange times. The lifetime of hard corona is long so it describes the biological characteristics of the particle (Monopoli et al. 2011).

Protein capping has been investigated on Au (Dobrovolskaia et al. 2009), CdSe/ZnS and Fe NPs (Rocker et al. 2009). The binding of protein has been estimated after incubation of 5–30 min, observing that protein adsorption to NPs surface is time dependent. Proteins with highest mobility initially adsorb to surface and then are replaced by high affinity and less mobile proteins. It may need several hours to complete this process. Slack and Horbett investigated that this process includes competitive interaction and binding of complex mixture of protein (about 3700) for a specific number of surface sites (Slack and Horbett 1995).

In almost all cases proteins present in high concentration initially adsorb on the surface, with the passage of time dissociate and replaced by less abundant and high affinity proteins. Surface properties, particle material and size are very important in determining the corona composition (Monopoli et al. 2011; Vroman et al. 1980). Understanding the specific time scales that represent the exchange of adsorbed proteins with free proteins and time evolution of capping proteins in a provided biological environment, making it possible to predict fate of particle regarded to its interactions with the particular receptors or its existence time in blood. Such understanding will help to comprehend the impact on nanomedicine and on nanotoxicity (Monopoli et al. 2011) (Fig. 6.4). The alteration in protein corona composition due to association/dissociation is termed as the “Vroman effect.” This effect shows that the nature of binding proteins changes with time.

The protein capping is dynamic in nature; with the passage of time there is continuous process of association and dissociation of proteins taking place due to which its composition never remains same. Ultimately the concentrations of adsorbed and free proteins becomes equal, association and dissociation rates for proteins also become equivalent hence, equilibrium is established and corona composition never disturbed due to constant exchange. But, if the NPs move to another biological fluid the composition of corona will change again (Walkey and Chan 2012).

During protein capping, proteins having highest association rates initially bind to inorganic nanomaterial. If the residence times of these proteins is short, they quickly dissociates, later replaced by other proteins with minimum association rates and maximum residence times. During adsorption of plasma protein, the Vroman effect has been divided into two stages “early” and “late” stage. During early stage it is observed that fibrinogen, albumin and IgG adsorb rapidly on inorganic surface, later replace by apolipoproteins and coagulation factors (Goppert and Muller 2005b). It is proposed by Mathematical modelings that albumin and fibrinogen with rapid dissociation rate and highest concentration arrive first for binding and substitute by apolipoproteins due to its less concentration and slow dissociation rate. For every nanomaterial early stage is not observed but the late stage of the Vroman effect

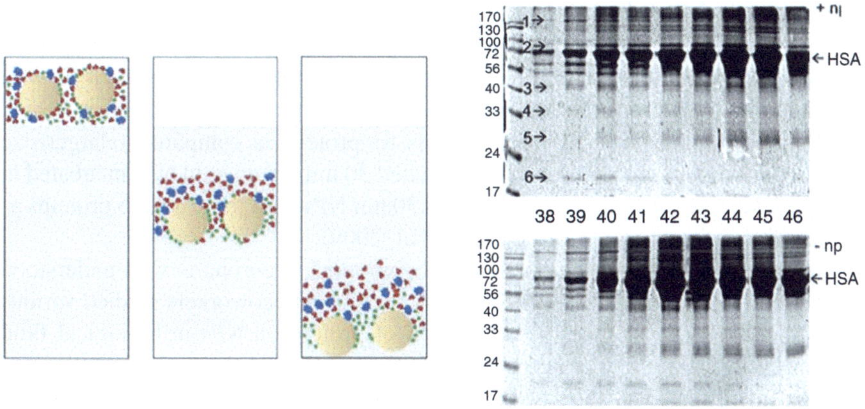

Fig. 6.4 Chromatographic study of size exclusion of NP-protein interactions. The elution time of proteins depends on the NP-protein interactions; proteins with high affinity for nanoparticles bound to NPs for longer time and elutes from the column earlier. Many different proteins collected from the size-exclusion column, are further subjected to gel electrophoresis for separation by using denaturing acrylamide gels. Mass spectrometry is used to identify the proteins from different gel bands. (Reproduced with permission from Ref. Cedervall et al. 2007a)

observed as low affinity proteins are replaced by high affinity proteins. The highest affinity of apolipoproteins is the main reason of adsorption on hydrophobic surface in replacement of fibrinogens (Goppert and Muller 2005a).

6.6 Effect of Different Parameters on the Composition of Capping Protein

There are many factors which are responsible for the interaction of NPs with biomolecules and composition of capping proteins. Biological environment and physicochemical properties of nanomaterial are very important parameters which are responsible for protein capping. Therefore, it is necessary to investigate and considerate these parameters for the safer plan of nanoparticles. Composition of nanoparticles and their surface properties are essential aspect in determining the identities and affinities of proteins that adsorb to NPs. Deng and coworkers investigated adsorption of available proteins to metal oxide nanoparticles such as zinc oxide, titanium dioxide and silicon dioxide having similar surface charges (Deng et al. 2009). The authors observed similar adsorb proteins on the surface of silicon dioxide and titanium NPs, while totally different proteins were examined on zinc oxide NPs. Alpha-2-acid glycoprotein, clusterin and apolipoprotein D were found in the corona of silicon dioxide and titanium NPs whereas corona of zinc oxide NPs was not composed of such proteins. Interestingly, corona of zinc oxide NPs contains haptoglobin (alpha), transferrin and Ig heavy chain alpha.

6.6.1 Effect of Nanoparticle Size

It was noted that in addition to proteins adsorption affinity, same nanoparticles with different sizes contain different composition of protein corona. It was investigated that NPs with smaller size have great affinity for proteins as compared to larger size nanoparticles. Dobrovolskaia et al. investigated 30 nm 50 nm gold NPs incubated in human plasma, interestingly they observed 30 nm NPs with more adsorb proteins as compared to 50 nm NPs (Dobrovolskaia et al. 2009).

The alteration in composition of adsorbs proteins in corona is well understood when protein size is near to the size of NPs. Tenzer and coworkers studied various sizes silica NPs (Si NP) for the adsorption of protein on NPs in biological fluid (Tenzer et al. 2011). It was noted that adsorption of 37% of the known proteins considerably affected by the size of SiNP. Interestingly, a minor change in particle size even 10 nm greatly affect the corona composition. Therefore, lipoprotein clust-sacaerin adsorb on the surface of small size SiNP whereas the actin regulatory protein gelsolin or prothrombin bound to the large size SiNP. Though, no relationship is found for the binding of some proteins like actin or immunoglobulin with the size of nanoparticles.

In a current study, the influence of NP size on hill constant and binding constant was investigated (Lacerda et al. 2009). The authors investigated gold NPs of a range of sizes (5–100 nm) incubated in biological fluid that contains albumin, histone, insulin, fibrinogen and γ-globulin. It was examined that the NPs size was drastically associated to the binding constant, K, in addition correlate to the extent of cooperatively of protein particle binding (Hill constant, n). It was observed that by increasing NP size binding constant also starts increasing while Hill constant start decreasing. As well as, they have also comprehended that by increasing NP size the amount of protein corona gradually starts increasing. In addition, conformational change also enhances with NP size.

In a study it was observed that NP size is responsible for the type of protein adsorb to form corona as well composition of the corona layer (Lundqvist et al. 2008). For example, apolipoproteinm B-100 adsorb to 100 nm NPs other than 50 nm NPs. Vertegel and co-workers examined effect of silica NPs size on adsorption of lysozyme on silica NPs (Vertegel et al. 2004). It was predicted that after adsorption of lysozyme on nanomaterial there is more protein unfolding on 100 nm NP as compared to 4 nm NP. It is concluded that particle size is vital aspect for determining structure and function of adsorb corona of nanoparticles.

6.6.2 Effect of Nanoparticle Shape

The adsorption of proteins on NP surface and the biological response to NP is sturdily governed by NP shape. For example, interactions of gold NPs with cell are influenced by shape of NPs. Like rod shaped gold NPs have less association in cell while spherical gold NP has more association and good interaction with cell (Cox et al. 1999). Deng and co-workers examined the effect of shape of titanium dioxide NPs on adsorption of protein (Deng et al. 2009). The authors reported that only spherical NPs adsorb proteins such as clusterin and apolipoprotein D as compared to nanotubes sand nanorods.

6.6.3 *Effect of Surface Functional Group and Coating*

In order to control the composition of capping proteins and to avert the protein adsorption, different groups can be used to functionalize the NP surface. It provides a "stealth character" to NP surface; consequently escape from being recognized by immune cells. Polyethylene glycol is a polymer that can be used to coat NP surface to reduce the adsorption of proteins and avoid their identity from RES (reticuloendothelial system) hence known as PEGylation. To facilitate the longer circulation time in blood the density of Polyethylene glycol on NPs surface can be controlled. Adsorption of protein can also be prevented by coating NPs surface with "Siliconate" (Perry et al. 2012; Gref et al. 2000). Polystyrene NPs in cultured endothelium cells were investigated in the presence of different functional groups (i.e., amidine, carboxyl, PEG, amine, methyl, cysteine and lysine) (Ehrenberg et al. 2009). It was predicted that adsorption ability of protein to the functionalized NPs surface revealed their affinity of interaction with cells. Moreover, interaction of NP with cell is not affected by the identity of adsorb proteins. Observations on poloxamine 908 coating polystyrene nanospheres showed a decrease in fibronectin binding in a comparison to uncoated nanospheres (Moghimi et al. 1993). Coating of Pluronic F127 on single walled carbon nanotubes and amorphous silica particles revealed an increase in the scattering of the NPs however binding of available proteins started decreasing (Dutta et al. 2007). Aggarwal et al. examined influence of many different coatings (i.e. PEG, Pluronic F127, dextran, polysorbate, poloxamer, poly (oxyethylene) and poloxamine) on biodistribution and adsorption of proteins on NPs and summarized the results (Aggarwal et al. 2009).

6.6.4 *Effect of Surface Charge*

The presence of surface charge on nanoparticles plays essential role in determining the composition of capping proteins and necessarily its ultimate fate in the biological environment. It is observed that increase in surface charge of NPs my increase the adsorption of proteins. Opsonins recognize the positively charged NPs as a result these particles are eliminated by the reticuloendothelial system (RES) and its final concentration in spleen and liver (Mahmoudi et al. 2011a, b, c, d; Aggarwal et al. 2009). To avoid opsonization, negatively charged groups can be used to coat surface of NPs that resulted in the formation of –ve zeta potential (30–50 mV). As the surface-coated NPs incubated in biological fluid, the surface binding proteins assist in huge reduction of their zeta potential 5–10 mV negative. Gessner et al. studied that adsorption of plasma proteins enhances by increasing surface charge density of nanoparticles (Gessner et al. 2002). Several investigations on polystyrene nanoparticles revealed that proteins with isoelectric points (PI) <5.5 binds with positively charged particles like albumin while proteins with isoelectric points >5.5 adsorb on negatively charged particles like IgG. Surface charge may also cause denaturation of binding proteins. A current study on gold NPs with negative, positive, and neutral ligands revealed that either negative or positive ligands cause denaturation of protein while natural structure of protein retained in the presence of neutral ligands (Lynch and Dawson 2008).

6.6.5 Effect of Hydrophilicity/Hydrophobicity

Hydrophobicity greatly affect the composition of capping protein and amount of adsorb protein. Adsorption of proteins is much more on hydrophobic surface of NPs as compared to hydrophilic surfaces. Many proteins that adsorb on the surface of hydrophobic NPs didn't retain their native structure and undergo protein denaturation (Roach et al. 2005). Similarly, apolipoproteins adsorb on the surface of hydrophobic NPs while IgG, albumin and fibrinogen becomes a part of protein corona on hydrophilic NPs (Cedervall et al. 2007a; Gessner et al. 2000). Cedervall et al. engaged ITC to examine the stoichiometry and adsorption affinity of proteins (Cedervall et al. 2007a, b). The authors observed that enhancement in hydrophobicity causes an increase in stoichiometry of proteins. Moreover at equilibrium point hydrophobic particles seem to be highly covered.

6.6.6 Effect of Biological Environment

Characteristics of NP are important but the composition of the natural medium in which NPs interact is a basic component in deciding the organization of protein corona (Nel et al. 2009). Maiorano et al. investigated the effect of composition of media on the development of protein corona (Maiorano et al. 2010). The authors investigated different estimated citrate-capped gold NPs and incubated them in cellular media like Roswell Park Memorial Institute medium (RPMI) and Dulbecco Modified Eagle's Medium (DMEM) that contains fetal bovine serum (FBS). These are the normally utilized cellular media and they vary from each other in glucose, amino acids and salt composition. Several techniques (plasmon resonance light scattering, dynamic light scattering and UV-visible) were utilized to estimate the protein capping on gold NPs intercede by RPMI and DMEM. It was observed that development of protein corona by using DMEM considerably depends upon time, whereas utilizing RPMI causes reduction of corona proteins. Characterization of Protein-NP complexes were also done by using mass spectroscopy and sodium dodecyl sulfate polyacrylamide gel electrophoresis (SDS-PAGE) and it was also observed that composition of protein corona never relies on the abundance of serum proteins. Viability assays were executed by using gold NPs of 15 nm on two cell lines U937 (human leukemic monocyte lymphoma cell line) and HeLa (human epithelial cervical cancer cell line) in both cellular media RPMI and DMEM. Fascinatingly, considerable differences in cellular uptake, biodistribution and dynamics were observed. More particularly, internalization of NPs with capping proteins in cells is prominently higher in RPMI media then DMEM media, bringing about higher Cytotoxicity impacts. In addition, the formation of protein corona is richer and stable in DMEM as compared to formation of protein corona in RPMI. These distinctions in the protein-NP complexes formed in different environments can influence the cellular response. So, aside from nanoparticles characterization, to understand the succeeding interaction of cell culture media with NPs, its evaluation must be done. Albanese et al. researched the conditioning of culture medium and its impact on the natural characteristics of NPs (Albanese et al. 2014). Absolute growth medium utilized for in vitro cell culturing usually contains serum that have different quantity of proteins or peptides. These proteins or peptides have the tendency for the formation of corona on NPs surface, in this way adjusting their cellular interaction and physicochemical properties. On the other hand, when absolute growth medium is used to culture the cells, protein or peptides composition with some other components may modify after some time because of cells metabolic activity, consequently influencing the formation of corona on NPs. Conditioning of cell culture medium means incubation of cells for a specific period of time and introduction of NPs in that medium, consequently contain by-products of cellular metabolic activity. From the research, it was investigated that conditioning of cell causes gold NP to form aggregates, which as a result change protein corona composition that depends on surface chemistry, NP size and cell phenotype. Additionally, it was shown that dynamic extracellular environment may modify the initial natural characters and as a result the cellular uptake.

6.6.7 Effect of Nanoparticles Material

Nano-sized silica and single-walled carbon nanotubes (SWCNT) were studied in the presence of plasma proteins exhibited different patterns of protein adsorption. SWCNT were seen to be coated with most abundant protein serum albumin while not found on silica NP. Nanoparticles with same surface charge such as SiO_2, ZnO and TiO_2 contains different plasma proteins (Table 6.3) (Deng et al. 2009).

Table 6.3 Identification of proteins bound to nanoparticles by mass spectrometry and gel electrophoresis

Nanoparticles	Proteins
TiO_2	Albumin, fibrinogen (α and β chains), histidine-rich glycoprotein, kininogen-1, complement C9 and C1q, Ig heavy chain (γ), fetuin A, vitronectin, apolipoprotein A1
SiO_2	Albumin, fibrinogen (α, β and γ chains), complement C8, Ig heavy chain (gamma, kappa), apolipoprotein A
ZnO	Albumin, Ig heavy chain (alpha, mu, gamma), apolipoprotein A1, immunoglobulin (J chain), alpha-2-macroglobulin, transferrin, alpha-1-antichymotryspin

6.7 Protein Stability/Activity Increased Upon Adsorption to Nanoparticles

In general, it is considered a drawback or a possible cause of NPs toxicity when a protein binding to nanoparticles loses its secondary structure as well as its natural activity, but it contains some positive outcome as well. NPs increase the stability of protein toward enzyme degradation; consequently increase the activity of enzymes by immobilization at surfaces. Enzymes for example *Pseudomonas cepacia* lipase (PCL) and *Candida rugosa* lipase (CRL) bind to polymethylmethacrylate (PMMA) and polystyrene NPs (PS) by simply adding lipase solution under protein-accommodating conditions (pH 7.6) to the polymeric NPs. Adsorption effect the selectivity and activity by enhancing the performance concerning those lipases bound to the same non-nanostructured bearers, and also improved the thermal stability and pH (Palocci et al. 2007).

It is observed that enzyme stability is also increased in the presence of much curved surface of C60 fullerenes in strongly denaturing environment. In an investigation it is observed that the enzyme soybean peroxidase exhibit the half life of ~2.5-fold higher when binds to fullerenes at 95 °C in comparison to enzyme that binds to graphite flakes and exhibit ~13-fold more higher half life as compared to native enzyme (Asuri et al. 2007a, b). Comparable perceptions with other nanostructures such as Au and silica nanoparticles have also been discussed. Sanghi et al. analyzed the capping protein around the silver nanoparticles to understand the stabilization of NPs by capping agents. Analysis was done by using SDS-PAGE which shows the presence of capping proteins which are responsible for stability of silver nanoparticles (Fig. 6.5). This capability to increase the protein stability affects several fields such as drug delivery, nanocomposites, diagnostics and sensors etc. (Sanghi and Verma 2009).

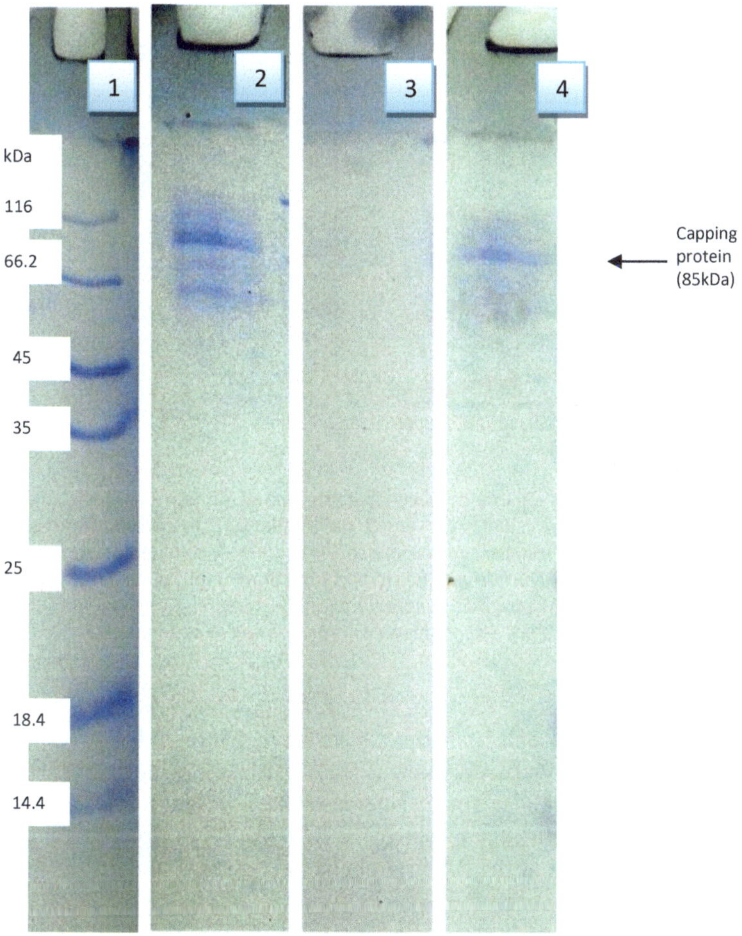

Fig. 6.5 Investigation of capping protein around the silver NPs by using SDS-PAGE. Lane *1*, molecular size marker; lane *2*, extracellular proteins in the cell filtrate; lane *3*, NPs loaded without boiling show no protein band; and lane *4*, NPs show a major 85-kDa capping protein which are loaded after boiling with 1% SDS loading buffer

6.8 Role of Nanoparticles in Fibrillation of Protein

NPs plays significant role in protein fibrillation by inducing conformational changes in proteins. There is a group of proteins called amyloidogenic proteins that aggregate in some conditions to form extremely insoluble structures (fibrils), which form plaques by forming precipitate. These aggregates are associated in a series of irreversible and severe pathological conditions (protein fibrillation or misfolding diseases), for example Parkinson's disease, Alzheimer's disease, and dialysis-related

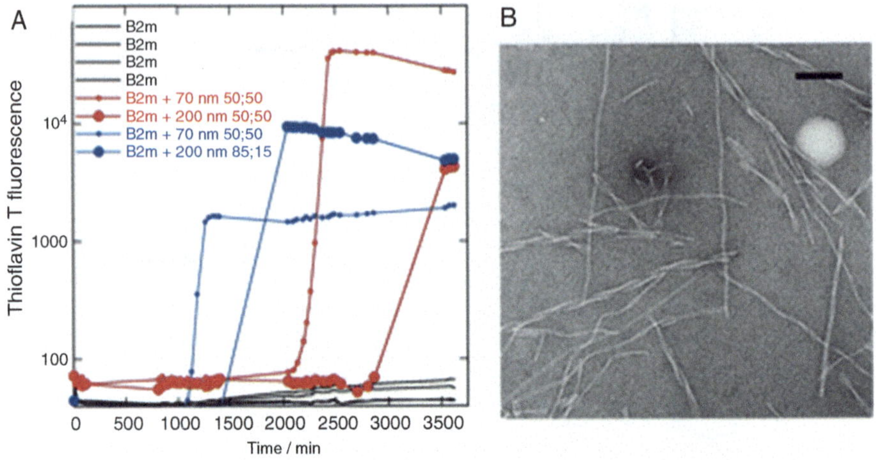

Fig. 6.6 Illustration of nanoparticles dependent fibrillation of amyloidogenic proteins, NPs increase the fibrillation by using assay based on thioflavin-T to protein fibril binding. (**a**) Thioflavin-T assays in the presence and absence (black) of NPs of different composition and size. When the thioflavin-T bound to fibrils, it fluoresces; the initiation of fluorescence depends on the onset of fibrillation. (**b**) TEM image of protein fibril with NPs representing that fibril do not move out from NPs. Scale bar: 100 nm. (Reproduced with permission from Linse et al. 2007)

amyloidosis. Minimum such 20 diseases have been identified. It is investigated that a range of different NPs (copolymer, cerium oxide, (PEG)-coated QDs, carbon nanotubes, and poly (ethylene glycol)) increased the fibrillation of β2-microglobulin (a amyloidogenic protein) due to enhanced localization of proteins on surface of NPs when protein exist at pH 2.5 in globular and in little molten state (Fig. 6.6) (Linse et al. 2007).

6.9 Use of Nanoparticles to Investigate Protein-Protein Interactions

Proteins interact with some new molecules to execute their biological function. To perform their function, enzymes, transcription factors and receptors must bind to their substrates, target DNA elements and ligands respectively. Therefore, any minute modification in protein conformation, after association with the surface of nanoparticle, significantly affects the function of protein and its interaction with other proteins (protein-protein interactions). In order to determine the biological effects of NPs, it is necessary to recognize the proteins that occupy the corona of nanoparticles and to find out the impact of integrated corona protein on the protein-protein interactions. Surface of NPs provides a new direction to investigate the protein interactions both protein-protein and protein-ligand. For instance, authors have examined adsorption of HSA (human serum albumin) in the abundance and deficiency of oleic acid with polymeric NPs of increasing hydrophobicity, which is key ligand that HSA gets adsorbed during its usual execution (Lindman et al. 2007). In the presence and absence of oleic acid extremely different interaction patterns were examined. HSA–oleic acid complex provides lower stoichiometry and an endothermic signal in comparison to the absence of oleic acid that provides exothermic signals. In an investigation it is observed that semiconductor nanocrystals (QDs with average diameter of less than 2 nm) form conjugates with HSA and can be utilized as fluorescent biological labels. Fluorescence resonance energy transfer (FRET) measurements from the amino acid tryptophan (Trp214) to the semiconductor nanocrystals, make it feasible to pursue the confined and overall changes in the structure HSA during thermal refolding and unfolding processes (Sarkar et al. 2007). It is clear from investigations that interaction of NPs to proteins is a dynamic process but still there is much to investigate.

6.10 Effect of Ignored Factors

Biological environment and characteristics of nanoparticles are significant factors in determining the composition of capping proteins as well as cellular response, but there are some other important hidden factors that play an essential role at the bio-nano interface. Recently, researches have been paying attention to comprehend these parameters known as 'ignored issues' which includes temperature, cell observer, gradient plasma, cell membrane composition and plasma concentration. It needs a comprehensive research to asses these ignored issues to facilitate the development of efficient and improved nanomedicine.

6.10.1 Effect of Plasma Concentration

The effect of plasma concentration on the composition of capping protein was studied by Monopoli et al. (2011). To observe the corona proteins and protein adsorption they used hydrophilic silica (SiO_2) and hydrophobic carboxylated polystyrene ($PSOSO_3$) NPs. Characterization of NP-protein complex was done by using DLS, zeta potential and DCS while electrospray liquid chromatography mass spectrometry (LC MS/MS) and 1D PAGE was used to determine hard corona composition semi quantitatively. An increase in the thickness of hard protein corona was observed by enhancing the concentration of plasma as shown in Fig. 6.7. Moreover, it was found that after partition from excess plasma, structure of NP-protein complexes *in vitro* and *in situ* is approximately same. In particular, the observation that hard corona composition depends upon protein concentrations will have critical effect when research performed on *in vitro* cell culture conditions were utilized to extrapolate over to *in vivo* conditions.

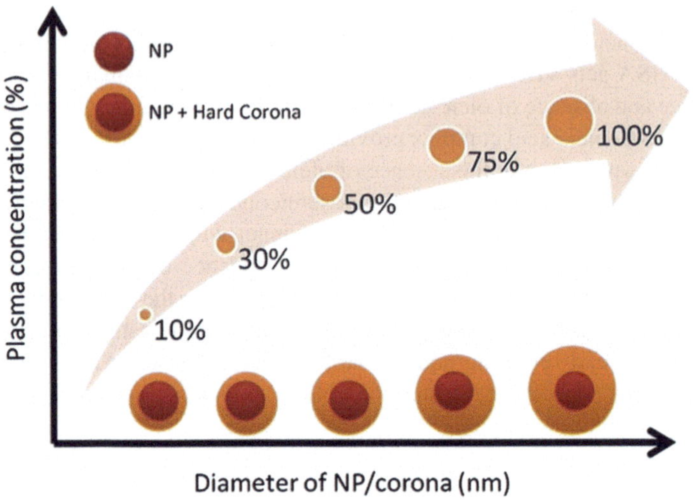

Fig. 6.7 Graphical illustration of effect of plasma concentration on the formation of thick hard corona on nanoparticles. The thickness of protein corona increases by increasing the plasma concentration

6.10.2 Temperature

A very important but not investigated factor that influences the composition of capping proteins is minor change in incubation temperature of NPs. It is understood that temperatures of the human body changes in different parts of the body (Rhoades and Pflanzer 1989), according to gender, and bodily activities. Body temperature drops to 28 °C when its peripheral parts are present in cold weather. It is observed that even intracellular living cells temperature is not homogeneous (Yang et al. 2011; Petersdorf 1974). Temperature of female body is slightly higher than the body temperature of male, and it can be influenced by the female's hormonal cycle. Body temperature decreases during sleep and physical activities raise the temperature upto 2 °C. In addition, body temperature increase to 41 °C in fever (Hasday and Singh 2000). Thus, it is confirmed that temperature of human body varies in the range of 35–39 °C. Mahmoudi et al. investigated the impact of temperature variation on the development and composition of capping proteins (Mahmoudi et al. 2013). Negatively charged polymer-coated FePt and fluorescently labeled NPs were incubated with human serum albumin (HSA) and to quantify the adsorb proteins at different temperatures and concentrations fluorescence correlation spectroscopy (FCS) was used. Additionally, different surface charge containing Dextran-coated FeOx NPs were also incubated with FBS (fetal bovine serum) at different temperatures.

It was accomplished that changes in incubation temperature affect the composition of capping proteins as well as cellular uptake. Mahmoudi et al. examined gold nanorods and the impact of plasmonic heat induction on the composition of their protein corona (Mahmoudi et al. 2014a, b). They used different concentration of FBS with cetyl trimethyl ammonium bromide (CTAB)-stabilized gold nanorods and incubated them. Afterward, composition of capping proteins were investigated before and after exposure to plasmonic heating induced by continuous laser irradiation. Characterization of NP protein complex was done by using UV-vis absorbance spectroscopy, ζ potential, transmission electron microscopy, and LC-MS/MS analysis. The authors concluded that thermal heating and irradiation both cause a considerable change in the hard protein corona composition whereas no change was observed on the surface charge of protein corona. In addition, it is observed that by applying different form of heating such as conventional thermal heating and plasmonic heating (i.e., photoinduced) on gold nanorods, a significant difference in the composition of protein corona was detected. It is concluded that biological activity of NPs may also affected due to changes in the composition of protein corona because of photoinduced local heating. Therefore, these alterations that have a great impact on the biological fate of nanoparticles must be considered for application of NPs in different fields.

6.10.3 Gradient Plasma Effect and Administrative Route

After entrance of NPs in the body, before approaching the target cell they get in touch with a huge number of biological components in the body. NPs enter in the body through different pathways, so it is important to evaluate the composition of capping proteins in gradient plasma media to know what the cell sees inside the cell. The type and concentration of proteins vary when the NPs move to reach the proposed target site; it depends on the administrative route (e.g., intramuscular, subcutaneous, intraosseous, intradermal, inhalation, intralumbar, and intravenous). Consequently, the composition of capping proteins was also observed to be different (Ghavami et al. 2013; Rahman et al. 2013). For instance, in the case of intravenous route and inhalation, the variation in the composition of protein corona was examined. When NPs directed to move through inhalation route, they come in contact with different biological fluids containing lung cell barrier, consequently almost different plasma proteins accumulate in comparison to blood, in the case of administration of NPs through intravenous route.

It was concluded by Ghavami et al. that plasma protein gradient significantly affect the capping protein composition and biological activity of NPs inside the cell (Ghavami et al. 2013). To investigate the impact of plasma concentration gradient, hydrophilic silica (SiO_2) and hydrophobic carboxylated polystyrene ($PSOSO_3$) nanoparticles were used. Characterization of NPs associated with proteins was done by liquid chromatography mass spectrometry (LC-MS/MS), zeta potential, differential centrifugal sedimentation (DCS), one dimensional sodium dodecyl sulfate polyacrylamide gel electrophoresis (1D PAGE), dynamic light scattering (DLS), transmission electron microscopy (TEM) and differential centrifugal sedimentation (DCS). It was revealed that quantity and composition of proteins present in the hard corona differentiate in both non-gradient plasma media and gradient plasma media (Ghavami et al. 2013).

6.10.4 Effect of Cell Observer and Cell Membrane

"Cell observer" (i.e., cell types) is another significant factor to recognize the fate of NPs *in vivo*. Firstly NPs get in touch with cell membrane of the cell, due to the presence of different surface protein such as sugars and phospholipids; cell membrane is different for each cell type. Most distinction in cell membranes becomes a cause of distinct cellular uptake and toxicity mechanisms (Laurent et al. 2013; Laurent et al. 2012). "Cell observer" concept plays an important role to understand the quantity of NP needed per cell in medical application and associated toxicity. Laurent et al. verified the affect of "cell observer" on the toxicity and uptake of super paramagnetic iron oxide NPs (SPIONs) (Laurent et al. 2012). Many cell lines for example, Hela, Panc-1, Capan-2, and Jurkat cells were exposed to interact with SPIONs having distinct surface chemistries and different interaction of each cell line with NPs was observed. For instance, glial cells, brain derived neuronal and lung cells depicted high level of toxicity as compared to other cell types with same SPIONs which shows moderate toxic effects. Particularly, cell type is responsible for both the toxicity and cellular uptake of SPIONs. When NPs enter in biological medium, an interface between cell membrane and NPs is created due to adsorption of proteins on the surface of NPs (CM) (Walkey and Chan 2012). Hence, impact of cell membrane composition must be taken into account to know well about interactions at the bio-nano interface. For example, negatively charged cell membrane causes penetration and *in vivo* cellular uptake with greater efficiency of positively charged NPs in comparison to negatively charged NPs as depicted in Fig. 6.8. In particular, adsorption capacity of the negatively charged NPs to cell membrane is lower which

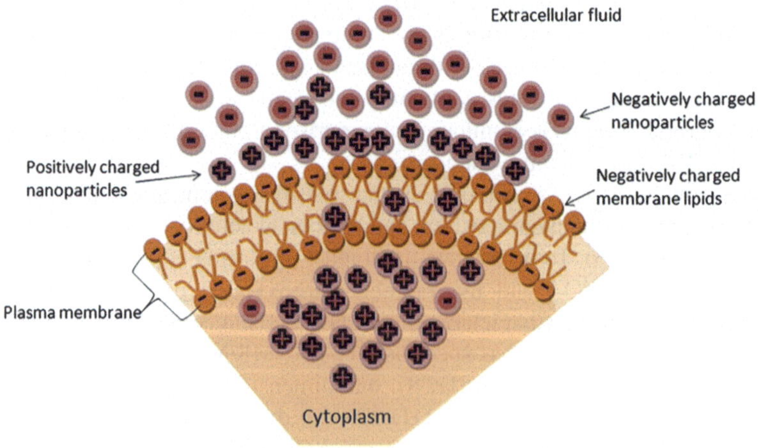

Fig. 6.8 Schematic representation of penetration and cellular internalization efficiency of positively charged nanoparticles as compared to negatively charged nanoparticles

leads in decreasing the possibility of cellular internalization. However, the cellular internalization of positively charged NPs can notably disturb the cell membrane and also enhances its toxicity (Mahmoudi et al. 2014a, b; Lesniak et al. 2013).

6.11 Conclusion

Small size of nanoparticles makes it easy to penetrate in any biological environment. After penetration of nanoparticles in a biological fluid, proteins compete to adsorb on the surface of NPs, the adsorption of a particular protein on the surface of nanoparticles formed nanoparticle-protein complexes that is usually considered as protein capping which composed of hard and a soft region, in hard region protein adsorb on NPs surface more strongly than soft region. The binding of proteins with nanoparticles may cause a change in the innate properties of both proteins and nanoparticles. Generally this protein capping is responsible for biological reactivity of nanoparticles. Various parameters and effect of several ignored issues have also been discussed which must be considered in future.

References

Aggarwal P, Hall JB, McLeland CB, Dobrovolskaia MA, McNeil SE (2009) Nanoparticle interaction with plasma proteins as it relates to particle biodistribution, biocompatibility and therapeutic efficacy. Adv Drug Deliv Rev 61(6):428–437

Albanese A, Walkey CD, Olsen JB, Guo H, Emili A, Chan WCW (2014) Secreted biomolecules alter the biological identity and cellular interactions of nanoparticles. ACS Nanoparticles 8(6):5515–5526

Asuri P, Bale SS, Pangule RC, Shah DA, Kane RS, Dordick JS (2007a) Structure, function, and stability of enzymes covalently attached to single-walled carbon nanotubes. Langmuir 23:12318–12321

Asuri P, Iseult L, Kenneth AD (2007b) Protein nanoparticles interaction. J Nanosci Nanotechnol 7:1675

Bardhan M, Mandal G, Ganguly T (2009) Steady state, time resolved, and circular dichroism spectroscopic studies to reveal the nature of interactions of zinc oxide nanoparticles with transport protein bovine serum albumin and to monitor the possible protein conformational changes. J Appl Phys 106:034701

Brandes N, Welzel PB, Werner C, Kroh LW (2006) Adsorption induced conformational changes of proteins onto ceramic particles, differential scanning calorimetry and FTIR analysis. J Colloid Interface Sci 299:56–69

Casals E, Pfaller T, Duschl A, Oostingh GJ, Puntes V (2010) Time evolution of the nanoparticle protein corona. ACS Nano 4:3623–3632

Cedervall T, Lynch I, Foy M, Berggad T, Donnelly S, Cagney G, Linse S, Dawson K (2007a) Detailed identification of plasma proteins adsorbed on copolymer nanoparticles. Angew Chem Int Ed 46:5754–5756

Cedervall T, Lynch I, Lindman S, Berggard T, Thulin E, Nilsson H (2007b) Understanding the nanoparticle-protein corona using methods to quantify exchange rates and affinities of proteins for nanoparticles. Proc Natl Acad Sci 104(7):2050–2055

Chakraborti S, Chatterjee T, Joshi P, Poddar A, Bhattacharyya B, Singh SP, Gupta V, Chakrabarti P (2009) Structure and activity of lysozyme on binding to ZnO nanoparticles. Langmuir 26:3506–3513

Chatterjee T, Chakraborti S, Joshi P, Singh SP, Gupta V, Chakrabarti P (2010) The effect of zinc oxide nanoparticles on the structure of the periplasmic domain of the *Vibrio cholerae* ToxR protein. FEBS J 277:4184–4194

Cox MC, Barnham KJ, Frenkiel TA, Hoeschele JD, Mason AB, He QY (1999) Identification of platination sites on human serum transferrin using 13C and 15N NMR spectroscopy. JBIC J Biol Inorg Chem 4(5):621–631

Deng ZJ, Mortimer G, Schiller T, Musumeci A, Martin D, Minchin RF (2009) Differential plasma protein binding to metal oxide nanoparticles. Nanotechnology 20:455101

Deng ZJ, Liang M, Monteiro M, Toth I, Minchin RF (2011) Nanoparticle-induced unfolding of fibrinogen promotes Mac-1 receptor activation and inflammation. Nat Nanotechnol 6:39–44

Deng ZJ, Liang M, Toth I, Monteiro M, Minchin RF (2013) Plasma protein binding of positively and negatively charged polymer-coated gold nanoparticles elicits different biological responses. Nanotoxicology 7:314–322

Ditto AJ, Shah PN, Yun YH (2009) Non-viral gene delivery using nanoparticles. Expert Opin Drug Deliv 6(11):1149–1160

Dobrovolskaia MA, Patri AK, Zheng J, Clogston JD, Ayub N, Aggarwal P, Neun BW, Hall JB, McNeil SE (2009) Interaction of colloidal gold nanoparticles with human blood, effects on particle size and analysis of plasma protein binding profiles. Nanomedicine 5:106–117

Dutta D, Sundaram SK, Teeguarden JG, Riley BJ, Fifield LS, Jacobs JM (2007) Adsorbed proteins influence the biological activity and molecular targeting of nanomaterials. Toxicol Sci 100(1):303–315

Eckhardt S, Brunetto PS, Gagnon J, Priebe M, Giese B, Fromm KM (2013) Nanobio silver: it's interactions with peptides and bacteria, and its uses in medicine. Chem Rev 113:4708–4754

Ehrenberg MS, Friedman AE, Finkelstein JN, Oberdorster G, McGrath JL (2009) The influence of protein adsorption on nanoparticle association with cultured endothelial cells. Biomaterials 30:603–610

Gasser M, Rothen-Rutishauser B, Krug HF, Gehr P, Nelle M, Yan B, Wick P (2010) The adsorption of biomolecules to multi-walled carbon nanotubes is influenced by both pulmonary surfactant lipids and surface chemistry. J Nanobiotech 8:1477–3155

Gessner A, Waicz R, Lieske A, Paulke B, Mader K, Muller RH (2000) Nanoparticles with decreasing surface hydrophobicities, influence on plasma protein adsorption. Int J Pharm 196(2):245–249

Gessner A, Lieske A, Paulke BR, Müller RH (2002) Influence of surface charge density on protein adsorption on polymeric nanoparticles: analysis by two-dimensional electrophoresis. Eur J Pharm Biopharm 54(2):165–170

Gessner A, Lieske A, Paulke B, Muller R (2003) Functional groups on polystyrene model nanoparticles. Influence on protein adsorption. J Biomed Mater Res A 65:319–326

Ghavami M, Saffar S, Abd Emamy B, Peirovi A, Shokrgozar MA, Serpooshan V (2013) Plasma concentration gradient influences the protein corona decoration on nanoparticles. RSC Adv 3(4):1119–1126

Gheshlaghi ZN, Riazi GH, Ahmadian S, Ghafari M, Mahinpour R (2008) Toxicity and interaction of titanium dioxide nanoparticles with microtubule protein. Acta Biochim Biophys Sin 40:777–782

Goppert TM, Muller RH (2005a) Adsorption kinetics of plasma proteins on solid lipid nanoparticles for drug targeting. Int J Pharm 302:172–186

Goppert TM, Muller RH (2005b) Polysorbate-stabilized solid lipid nanoparticles as colloidal carriers for intravenous targeting of drugs to the brain, comparison of plasma protein adsorption patterns. J Drug Target 13:179–187

Gref R, Luck M, Quellec P, Marchand M, Dellacherie E, Harnisch S (2000) Stealth' corona-core nanoparticles surface modified by polyethylene glycol (PEG), influences of the corona (PEG chain length and surface density) and of the core composition on phagocytic uptake and plasma protein adsorption. Colloids Surf B Biointerf 18(3–4):301–313

Hasday JD, Singh IS (2000) Fever and the heat shock response: distinct, partially overlapping processes. Cell Stress Chaperones 5(5):471–480

Hellstrand E, Lynch I, Andersson A, Drakenberg T, Dahlback B, Dawson KA, Linse S, Cedervall T (2009) Complete high-density lipoproteins in nanoparticles corona. FEBS J 276:3372–3381

Hill HD, Millstone JE, Banholzer MJ, Mirkin CA (2009) The role radius of curvature plays in thiolated oligonucleotide loading on gold nanoparticles. ACS Nano 3(2):418–424

Hussain SM, Hess KL, Gearhart JM, Geiss KT, Schlager JJ (2005) *In vitro* toxicity of nanoparticles in BRL 3A rat liver cells. Toxicol In Vitro 19:975–983

Jansch M, Stumpf P, Graf C, Ruhl E, Müller RH (2012) Adsorption kinetics of plasma proteins on ultrasmall superparamagnetic iron oxide (USPIO) nanoparticles. Int J Pharm 428:125–133

Karajanagi SS, Vertegel AA, Kane RS, Dordick JS (2004) Structure and function of enzymes adsorbed onto single-walled carbon nanotubes. Langmuir 20:11594–11599

Kathiravan A, Paramaguru G, Renganathan R (2009) Study on the binding of colloidal zinc oxide nanoparticles with bovine serum albumin. J Mol Struct 934:129–137

Kircher MF, de la Zerda A, Jokerst JV, Zavaleta CL, Kempen PJ, Mittra E (2012) A brain tumor molecular imaging strategy using a new triple-modality MRI-photoacoustic-Raman nanoparticle. Nat Med 18(5):829–834

Lacerda SHDP, Park JJ, Meuse C, Pristinski D, Becker ML, Karim A, Douglas JF (2009) Interaction of gold nanoparticles with common human blood proteins. ACS Nano 4:365–379

Laurent S, Burtea C, Thirifays C, Hafeli UO, Mahmoudi M (2012) Crucial ignored parameters on nanotoxicology: the importance of toxicity assay modifications and "cell vision". PLoS One 7(1):29997

Laurent S, Burtea C, Thirifays C, Rezaee F, Mahmoudi M (2013) Significance of cell "observer" and protein source in nanobiosciences. J Colloid Interface Sci 392:431–445

Lee JH, Jang JT, Choi JS, Moon SH, Noh SH, Kim JW (2011) Exchange-coupled magnetic nanoparticles for efficient heat induction. Nat Nanotechnol 6(7):418–422

Lesniak A, Salvati A, Santos-Martinez MJ, Radomski MW, Dawson KA, Aberg C (2013) Nanoparticle adhesion to the cell membrane and its effect on nanoparticles uptake efficiency. J Am Chem Soc 135(4):1438–1444

Liang XJ, Meng H, Wang YZ, He HY, Meng J, Lu J (2010) Metallofullerene nanoparticles circumvent tumor resistance to cisplatin by reactivating endocytosis. Proc Natl Acad Sci 107(16):7449–7454

Lindman S, Lynch I, Thulin E, Nilsson H, Dawson KA, Linse S (2007) Systematic investigation of the thermodynamics of HSA adsorption to N iso propylacrylamide/N tert-butylacrylamide copolymer nanoparticles. Effects of particle size and hydrophobicity. Nano Lett 7:914

Linse S, Sara L, Celia C, Wei FX, Iseult L, Stina L, Eva T, Sheena ER, Kenneth AD (2007) Nucleation of protein fibrillation by nanoparticles. Proc Natl Acad Sci U S A 104:8691

Liu S, Sui Y, Guo K, Yin Z, Gao X (2012) Spectroscopic study on the interaction of pristine C60 and serum albumins in solution. Nanoscale Res Lett 7:433

Lundqvist M, Sethson I, Jonsson BH (2004) Protein adsorption onto silica nanoparticles. Conformational changes depend on the particles' curvature and the protein stability. Langmuir 20:10639 10647

Lundqvist M, Stigler J, Elia G, Lynch I, Cedervall T, Dawson KA (2008) Nanoparticle size and surface properties determine the protein corona with possible implications for biological impacts. Proc Natl Acad Sci 105:14265–14270

Lynch I, Dawson KA (2008) Protein-nanoparticle interactions. Nano Today 3(1–2).40–47

Mahmoudi M, Lynch I, Ejtehadi MR, Monopoli MP, Bombelli FB, Laurent S (2011a) Protein nanoparticle interactions, opportunities and challenges. Chem Rev 111(9):5610–5637

Mahmoudi M, Sahraian MA, Shokrgozar MA, Laurent S (2011b) Superparamagnetic iron oxide nanoparticles, promises for diagnosis and treatment of multiple sclerosis. ACS Chem Neurosci 2(3):118–140

Mahmoudi M, Sant S, Wang B, Laurent S, Sen T (2011c) Superparamagnetic iron oxide nanoparticles (SPIONs): development, surface modification and applications in chemotherapy. Adv Drug Deliv Rev 63(1–2):24–46

Mahmoudi M, Shokrgozar MA, Sardari S, Moghadam MK, Vali H, Laurent S, Stroeve P (2011d) Irreversible changes in protein conformation due to interaction with superparamagnetic iron oxide nanoparticles. Nanoscale 3:1127–1138

Mahmoudi M, Abdelmonem AM, Behzadi S, Clement JH, Dutz S, Ejtehadi MR (2013) Temperature: the "ignored" factor at the nanobio interface. ACS Nano 7(8):6555–6562

Mahmoudi M, Lohse SE, Murphy CJ, Fathizadeh A, Montazeri A, Suslick KS (2014a) Variation of protein corona composition of gold nanoparticles following plasmonic heating. Nano Lett 14(1):6–12

Mahmoudi M, Meng J, Xue X, Liang XJ, Rahman M, Pfeiffer C (2014b) Interaction of stable colloidal nanoparticles with cellular membranes. Biotechnol Adv 32(4):679–692

Maiorano G, Sabella S, Sorce B, Brunetti V, Malvindi MA, Cingolani R, Pompa PP (2010) Effects of cell culture media on the dynamic formation of protein – nanoparticle complexes and influence on the cellular response. ACS Nano 4:7481–7491

Miclaus T, Bochenkov VE, Ogaki R, Howard KA, Sutherl DS (2014) Spatial mapping and quantification of soft and hard protein coronas at silver nanocubes. Nano Lett 14(4):2086–2093

Mishra SR, Zhang B, Si Z, Yang L, Jiang W, Yan B (2009) Characterization of protein clusters of diverse magnetic nanoparticles and their dynamic interactions with human cells. J Phys Chem C 113:5390–5395

Moghimi SM, Muir I, Illum L, Davis SS, Kolb-Bachofen V (1993) Coating particles with a block co-polymer (poloxamine-908) suppresses opsonization but permits the activity of dysopsonins in the serum. Biochimica et Biophysica Acta (BBA) Mol Cell Res 1179(2):157–165

Monopoli MP, Walczyk D, Campbell A, Elia G, Lynch I, Bombelli FB, Dawson KA (2011) Physical-chemical aspects of protein corona, relevance to *in vitro and in vivo* biological impacts of nanoparticles. J Am Chem Soc 133:2525–2534

Nel AE, Madler L, Velegol D, Xia T, Hoek EM, Somasundaran P (2009) Understanding biophysicochemical interactions at the nano-bio interface. Nat Mater 8(7):543–557

Palocci C, Laura C, Iole V, Enrico C, Marco D, Ilaria F, Maria V (2007) Lipolytic enzymes with improved activity and selectivity upon adsorption on polymeric nanoparticles. Biomacromolecules 8:3047

Perry JL, Reuter KG, Kai MP, Herlihy KP, Jones SW, Luft JC (2012) PEGylated PRINT nanoparticles, the impact of PEG density on protein binding, macrophage association, biodistribution, and pharmacokinetics. Nano Lett 12(10):5304–5310

Petersdorf R (1974) Chap 12. Chills and fever. In: Wintrob MM, Thorn GW, Adams RD, Braaunwald E, Isselbacher KJ, Petersdorf RG (eds) Harrison's principles of internal medicine, 7th edn. McGraw-Hill, New York, pp 58–60

Ragnaill MN, Brown M, Bramini M, Callanan S, Lynch I (2011) Internal benchmarking of a human blood–brain barrier cell model for screening of nanoparticle uptake and transcytosis. Eur J Pharm Biopharm 77(3):360–367

Rahman M, Laurent S, Tawil N, Yahia L H, Mahmoudi M (2013). Nanoparticle and protein corona. In Protein-nanoparticle interactions. Springer, New York, 11:21–44

Rhoades R, Pflanzer RG (1989) Human physiology, 4th edn. Saunders College Publication, Fort Worth

Roach P, Farrar D, Perry CC (2005) Interpretation of protein adsorption: surfaceinduced conformational changes. J Am Chem Soc 127(22):8168–8173

Rocker C, Potzl M, Zhang F, Parak WJ, Nienhaus GU (2009) A quantitative fluorescence study of protein monolayer formation on colloidal nanoparticles. Nat Nanotechnol 4:577–580

Sanghi R, Verma P (2009) Biomimetic synthesis and characterization of protein capped silver nanoparticles. Bioresour Technol 100:501–530

Sarkar R, Shankara N, Lars OP, Fernando D, Andy M, Samir KP (2007) Direct conjugation of semiconductor Nanocrystals to a globular protein to study protein-folding intermediates. J Phys Chem B 111:12294
Simberg D, Park JH, Karmali PP, Zhang WM, Merkulov S, McCrae K (2009) Differential proteomics analysis of the surface heterogeneity of dextran iron oxide nanoparticles and the implications for their in vivo clearance. Biomaterials 30(23):3926–3933
Slack SM, Horbett TA (1995) The Vroman effect. ACS Symp Ser 602:112–128
Sund J, Alenius H, Vippola M, Savolainen K, Puustinen A (2011) Proteomic characterization of engineered nanomaterial–protein interactions in relation to surface reactivity. ACS Nano 5:4300–4309
Tanaka T, Mangala LS, Vivas-Mejia PE, Nieves-Alicea R, Mann AP, Mora E (2010) Sustained small interfering RNA delivery by mesoporous silicon particles. Cancer Res 70(9):3687–3696
Tenzer S, Docter D, Rosfa S, Wlodarski A, Kuharev J, Rekik A, Knauer SK, Bantz C, Nawroth T, Bier C (2011) Nanoparticle size is a critical physicochemical determinant of the human blood plasma corona: a comprehensive quantitative proteomic analysis. ACS Nano 5:7155–7167
Turci F, Ghibaudi E, Colonna M, Boscolo B, Fenoglio I, Fubini B (2010) An integrated approach to the study of the interaction between proteins and nanoparticles. Langmuir 26:8336–8346
Verma A, Stellacci F (2010) Effect of surface properties on nanoparticle–cell interactions. Nanoscale Res Lett 6:12–21
Vertegel AA, Siegel RW, Dordick JS (2004) Silica nanoparticle size influences the structure and enzymatic activity of adsorbed lysozyme. Langmuir 20(16):6800–6807
Vroman L (1962) Effect of adsorbed proteins on the wettability of hydrophilic and hydrophobic solids. Nature 196:476–477
Vroman L, Adams AL, Fischer GC, Munoz PC (1980) Interaction of high molecular-weight kininogen, factor-xii, and fibrinogen in plasma at interfaces. Blood 55:156–159
Walczyk D, Bombelli FB, Monopoli MP, Lynch I, Dawson KA (2010) What the cell "sees" in bionanoscience. J Am Chem Soc 132:5761–5768
Walkey CD, Chan WC (2012) Understanding and controlling the interaction of nanomaterials with proteins in a physiological environment. Chem Soc Rev 41:2780–2799
Wangoo N, Suri CR, Shekhawat G (2008) Interaction of gold nanoparticles with protein, a spectroscopic study to monitor protein conformational changes. Appl Phys Lett 92:133104
Word JM, Schiller T, Yan BJ (1999) Topological analysis and interactive visualization of biological networks and protein structures. Mol Biol 285:1711
Worrall JWE, Verma A, Yan HH, Rotello VM (2006) Cleaning of nanoparticle inhibitors via proteolysis of adsorbed proteins. Chem Commun 0:2338–2340
Yang JM, Yang H, Lin L (2011) Quantum dot nano thermometers reveal heterogeneous local thermogenesis in living cells. ACS Nano 5(6):5067–5071

Chapter 7
Nanobotany and Pharmaceuticals

Zunera Iqbal and Sumera Iqbal

7.1 Introduction

Bionanotechnology is the combination of biotechnology and nanotechnology and it emerged as an ecofriendly technology for the nanomaterials biological synthesis. A physicist Professor Richard Feynman in his remarkable speech "there's plenty of room at the bottom" devised the idea of nanotechnology (Feynman 1959), yet the Professor Norio Taniguchi of Tokyo Science University introduced the term nanotechnology (Taniguchi 1974). Nanoparticles are regarded as groups of elements in the range of 1–100 nm. These nanoparticles possess different kind of physiochemical properties; depending upon the material and method from which they are synthesized.

Nanobotany is the detailed study of different nanomaterial i.e., nanoparticles, nanostructures and nanodevices synthesized from any plant material. These synthesized nanomaterials have some kind of anti-microbial, anti-oxidant, anti-tumor anti-inflammatory and anti-cancer activities. Using plant material to synthesize these nanoparticles either enhances or decreases these activities of nanoparticles.

Due to the increase in drug resistance frequency among pathogenic bacteria and fungi, there is a need of developing new antimicrobials for these infectious diseases. Nanoparticles appear as one of the most favorable healing agents. Nanoparticles has distinctive physiochemical properties and against microbes has high growth inhibition capability, this directed to the rise in the study of application of nanoparticles as antimicrobials (Rai and Bai 2011).

Biosynthesis of nanoparticles emerged as a cost-effective and ecofriendly method as compared to physical and chemical methods. Plants act as bioreactors for the nanoparticles synthesis. These nanoparticles can be metal (silver, gold and platinum) or metal oxide nanoparticles (iron oxide, zinc oxide or aluminum oxide nanoparticle). Silver and zinc oxide nanoparticles have gain special attention as

Z. Iqbal · S. Iqbal (✉)
Department of Botany, Lahore College for Women University, Lahore, Pakistan

S. Javad, A. Butt (eds.), *Nanobotany*, https://doi.org/10.1007/978-3-319-77119-9_7

antimicrobial agent due to their unique properties and are extensively used in various biomedical applications and other industries.

Some of the applications of nanomaterials in the fields of bio-medical and agriculture are bio-detection of pathogens, gene and drug delivery, DNA structure probing, splitting and refining of biological cells and molecules, detection of proteins, improvement of MRI contrast, anti-microbial tumor destruction via heating (hyperthermia) and anti-insect molecules (Sahayaraj and Rajesh 2011). Antimicrobial assays are necessary to determine the antimicrobial potential of different nanoparticles.

7.2 Antimicrobial Assays

To check the different compounds' inhibition effects against pathogens, antimicrobial assays are done and then their (broad vs. narrow) inhibitory spectra are established. The understanding of their inhibitory spectra of antimicrobial agents is important, before their application in various fields. To evaluate the *in vitro* antimicrobial activity of a compound or pure extract, various laboratory methods can be used.

Due to the hold to the properties of new antimicrobial products like fighting the multidrug-resistant bacteria, it is important to develop a better understanding of the current methods available for screening and quantifying the antimicrobial effect of an extract or a pure compound for its applications in human health, agriculture and environment. The best recognized and simple methods are the agar well diffusion, the disc diffusion and agar dilution methods.

7.2.1 Agar Disc-Diffusion Method

This technique was established in 1940 and for the routine antimicrobial susceptibility testing, it is the approved method used by several scientific microbiology laboratories. Many conventional and appropriate standards have been published by the Clinical and Laboratory Standards Institute (CLSI) for yeasts and bacteria testing.

Although, by this method not all the pathogenic bacteria can be tested precisely, to access microbial pathogens like Haemophilus Para influenza, Haemophilus influenza and Streptococci the standardization has been made using particular culture media, several incubation conditions and criteria for inhibition zones (Balouri et al. 2016).

In this process, with a uniform inoculum of the test microorganism, agar plates are inoculum. Then on the agar surface, filter paper discs (about diameter of 6 mm) are placed containing the test compound (i.e., antibiotics or plant extract) at different concentrations. Under suitable conditions, petri dishes are incubated. The germination and growth of the test microorganisms inhibited usually after the diffusion of

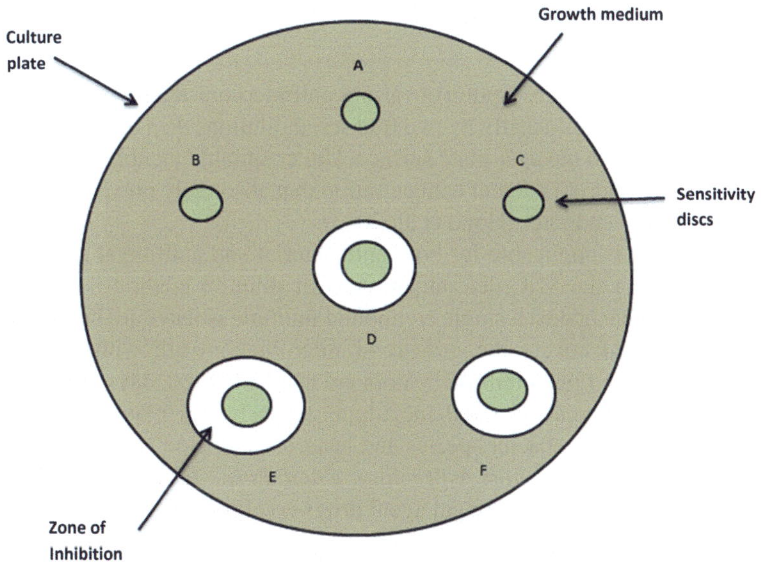

Fig. 7.1 Agar disc diffusion method. Bacteria are resistant to sensitivity discs *A*, *B* and *C* while are sensitive to discs *D*, *E* and *F* containing different concentrations of nanoparticles

antimicrobial agent in agar. Then the diameters of growth inhibition zones are measured (Fig. 7.1).

This process cannot distinguish the bactericidal and bacteriostatic effect as bacterial growth inhibition does not specify the bacterial death. Also the minimum inhibitory concentration (MIC) cannot be determined by this method as it is difficult to measure the quantity of extract (material) dispersed into the agar. Due to its easiness and less price this method is commonly used for antimicrobial analysis of drugs, essential oils and other plant extracts.

7.2.2 *Agar Well Diffusion Process*

This method is broadly used to estimate the antimicrobial potential of microbial agents or plants. At first, by spreading of the microbial inoculum, the entire agar plate surface is inoculated. Then, with a sterile cork borer, a diameter of 6–8 mm well is punched aseptically. Then at desired concentration, antimicrobial extract is added into the well. Depending upon the verified pathogens, under favorable conditions agar plates is incubated. By diffusing in the agar medium, the antimicrobial agent inhibits the growth of the microbial strain tested.

7.2.3 Agar Dilution Method

This method includes the absorption of various antimicrobial agents on agar medium at desired concentrations usually by two fold serial dilutions then a microbial inoculum is inoculated onto the agar plate surface. Under suitable incubation conditions, the minimum antimicrobial agent concentration that absolutely prevents the growth is recorded as MIC endpoint (Hood et al. 2003).

This procedure is applicable for both antibacterial and antifungal susceptibility testing. Usually for the MIC determination, agar dilution method is preferred to broth dilution, when against a single compound multiple isolates are being tested or if the extract tested covers the growth of microbial growth with its coloring. Commercially formed inoculum replicators are available these days that can transfer up to 32–60 different bacterial inoculums to each agar plate. For fastidious organisms such as Helicobacter species and anaerobes, this procedure is suggested as standardized method. Against *Aspergillus, Candida sp., Fusarium and dermatophytes*, it has been used as antifungal agent drugs combinations.

7.2.4 Broth Dilution Method

One of the basic antimicrobial susceptibility testing methods is Macrobroth and tube-dilution. In this process, two fold dilutions of antibiotics (1, 2, 4, 8 and 16 μg/mL are prepared in a liquid growth medium using test tubes. Then with a 1–5 × 10^5 CFU/mL standardized bacterial suspension, the antibiotic containing tubes were inoculated. The tubes were examined by turbidity revealing visible bacterial growth after overnight incubation at 35 °C (CLSI Methods 2012).

The last antibiotic concentration that prevented the growth considered as minimal inhibitory concentration (MIC). To get a quantitative result, this technique is useful. It has some disadvantages such as difficulty in the preparation of antibiotic solutions every time and probability of errors while forming dilutions, also huge space and quantity of chemicals are required for each test. 96-well tray is used for this assay. All these methods are used to determine the antimicrobial potential of nanoparticles.

7.3 Anti-microbial Properties of Metal Nanoparticles

Metallic nanoparticles emerged as strong antimicrobial agents in recent few years. Silver and gold nanoparticles have strong antimicrobial properties and they are extensively used in drug delivery and other biomedical applications. These nanoparticles have unique physiochemical properties due to which they can be manipulated in any shape and size according to the desired applications. (Makarov et al. 2014).

7.3.1 Antimicrobial Properties of Silver Nanoparticles

Against a wide range of microorganisms form diverse classes such as gram negative and gram-positive bacteria, fungi and viruses, silver is considered as a well-known antimicrobial agent. In China, as antimicrobial agents in most of the public places such as railway stations and elevators, silver nanoparticles are used.

They are also used in surgically embedded tubes as antimicrobial agents to reduce the toxicities occurred during surgery and they also have anti-fungal, anti-angiogenic, anti-inflammtory and anti-permeability activities (Kalishwaralal et al. 2009; Gurunathan 2009; Sheikpranbabu et al. 2009). In ancient times, silver and silver nitrate has been broadly used for wound healing practices but the antimicrobial potential of silver metal increased hundred times by the silver nanoparticles. These nanoparticles can be synthesized using different methods and synthesis depends upon the reduction of silver ions. These silver ions interact with microorganisms (bacteria and fungi) when their antimicrobial potential is tested.

7.3.2 How Silver Nanoparticles Effect Bacterial Cells?

Silver nanoparticles contain silver ions that bind with the bacterial cell wall, form pits on cell membrane thus disturbing the cell permeability and destroy the structure of bacterial cell. Then these silver ions penetrate inside the cell, enhance the fabrication of (ROS) reactive oxygen species, causing the oxidative stress that ultimately lead to cellular inactivation. Silver nanoparticles also disturb the ribosomal machinery of bacterial cells; disturb the DNA replication by interacting with DNA bases that leads to cellular death.

It is supposed that silver atoms forms stable SAAg bonds with thiol containing compounds when bind to thiol groups and causes the deactivation of call membrane enzymes including ion transport and trans-membrane energy generation. Silver ions attach between the purine and pyrimidine base pairs after entering the cell and denature the DNA by disturbing the hydrogen bonding between the two anti-parallel strands (Ahmed et al. 2016) (Fig. 7.2).

7.3.3 Influence of Silver Nanoparticles on Gram Positive and Gram Negative Bacteria

The silver nanoparticles antimicrobial properties depend on:

1. Size and physiochemical conditions (pH, ionic strength).
2. Capping agent.

The Ag^{+} ions positive charge is proposed vital for antimicrobial activities. In ionized form silver is inactive but when it comes in contact with moisture it releases silver ions and show antimicrobial properties.

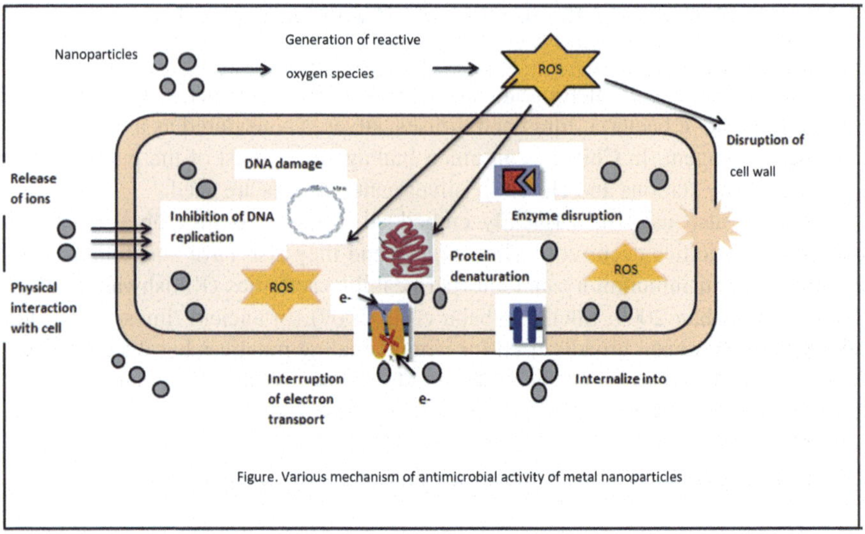

Fig. 7.2 Antimicrobial effect of metal nanoparticles on bacterial cells

In some literature it is shown that there must be some kind of contact between negatively charged bacterial cell walls and positively charged silver ions. The antimicrobial effect of AgNPs is dose dependent. Silver nanoparticles found to be accumulated in the *E. coli* bacterial cell membrane after the interaction with the bacterial cells. It results in the death of cell due to increase in cell membrane permeability.

Gram negative bacteria are more susceptible to Ag^+ than gram positive. Gram positive bacteria cell wall is thicker due to the presence of large number of peptidoglycan molecules than gram negative bacteria. The peptidoglycan call wall is negatively charged and it interacts more with the Ag^+ ions and may get stuck by peptidoglycan in gram positive bacteria (Ankana et al. 2010). The antibacterial properties of nanoparticles depend on their total surface area. Due to the large surface area to volume ratio, small nanoparticles have greater antibacterial activity.

When silver nanoparticles interact with DNA and enzymes of bacterial cells through an electron release mechanism (Sharma et al. 2009) or free radical production (Ankana et al. 2010) they causes the inactivation of enzymes and disruption of DNA molecule.

7.3.4 How Do Silver Nanoparticles Affect Fungi?

As an antimicrobial agent and sterilizer, silver has long history of use. It can interact with the protein contents of microorganisms by attracting disulphide bond thus change the three dimensional structure of proteins. It can stop the functional processes of microorganisms (Ahmed et al. 2016).

Against broad spectrum fungi such as *Aspergillus*, *Candida* and *Sacchromyces*, nanosilver act as much effective and fast acting fungicide. Small nanoparticles have high particle penetration in microbial cells therefore they exhibit higher antimicrobial activity than large size nanoparticles (Yu et al. 2005).

The effects of antifungal action of silver nanoparticles were investigated against sclerotium-forming phytopathogens especially against *Sclerotinia sclerotiorum* and *S. minor and Rhizoctonia solani* by Min and his fellows (Min et al. 2009).

The antimicrobial efficiency of the silver nanoparticles observed in fungi on their hyphal growth in the following order, *R. solani* > *S. sclerotiorum* > *S. minor*. At low concentrations of silver nanoparticles, the sclerotium development of S. sclerotiorum was successfully inhibited. The hyphae interacted with silver nanoparticles was highly damaged shown in a microscopic observation and layers of the hyphal wall get separated and breakdown the hyphae. It is suggested to use silver nanoparticles as pesticides for phyto pathogenic fungal controls.

The antibacterial activity of polymeric medical devices can be increased by silver nanoparticles used as impregnation. Surgical masks and implantable devices are silver impregnated medical devices and they have major antimicrobial efficacy (Furno et al. 2004).

7.3.5 Cytotoxicity of Silver Nanoparticles

It can be checked to determine anticancer potential of silver nanoparticles using MTT assay. MTT (3-(4,5-dimehylthiazol-2,5-diphenylformazan bromide) is a yellow color soluble tetrazolium salt that reduce into purple color formazan salt during this test. MTT is used to determine the cells viability. 96-well plate is used for this method. Different concentrations of silver nanoparticles are used to check their cytotoxicity potential against viable cancer cells. Dead cells do not reduce MTT tetrazolium salt.

1. Cells seeded into 96-well microtitre plate.
2. Different conc. of silver nanoparticles added in wells.
3. Then incubate it for 24–36 h at 37 °C.
4. After this, 50 μl of MTT is added into plates and is again incubated for 24 h.
5. Absorbance is measured by microtiter at Optical density 450 nm.
6. A graph of absorbance against concentration is plotted to check cytotoxicity of silver nanoparticles.

7.4 How Gold Nanoparticles Act as Antimicrobial Agent?

In the Chinese medical history in 2500 BC, the medicinal usage of gold can be found. In the Indian Ayurvedic medicine in the name of ('Swarna" means gold and 'Bhasma" means ash), red colloidal gold is still in use during old age for transformation and recovery (Higby 1990).

In the western world, gold has an extensive history of use as nervine, a substance used for the recovery of mental conditions. For the cure of epilepsy, gold was also recommended in the sixteenth century. During nineteenth century, gold was used for the cure of syphilis. For the tuberculosis, a therapy based on gold was introduced in 1920s after Robert Koch discovery of gold cyanide bacteriostatic effect against *Tubercle bacillus* (Shaw 1999). For the treatment of rheumatic diseases, gold compounds have most important medical uses.

Because of their biocompatibility, gold particles are briefly exploited in organisms (Bhattacharya and Mukherjee 2008). Biologically gold nanoparticles are considered to be inactive but to have their chemical or photothermal functionality, they can be engineered. To photo thermally kill *Staphylococcus aureus* by laser, light absorbing gold nanoparticles attached with specific antibodies have been used (Zharov 2006).

Gold NPs absorbs near infrared irradiation and can kill bacterial and cancer cells via thermal heating. It has been used as chemotherapeutic agent for various drug resistant microorganisms for example efficacy of vancomycin antibiotic enhanced when covering through gold NPs in opposition to Vancomycin-resistant enterococci (VRE). Reduced gold NPs with Cofactor have antimicrobial action in opposition toward Gram +ve and Gram -ve bacteria. Cofactor prevents the formation of peptidoglycan layer after that cell wall become porous then further gold NPs make holes in cell wall which tend to the discharge of substances of cell as well as death of cell (Ravishankar et al. 2011).

7.4.1 How Gold Nanoparticles Effect Bacterial Cells?

Gold nanoparticles prove to be as antifungal agent but they have antibacterial activity at high concentrations. Their mechanism of action is different for Gram negative and Gram positive bacteria due to their cell wall differences. Gram positive bacteria has thick layer of peptidoglycan thus they require high dose of nanoparticles.

Gold nanoparticles affects the membrane potential of bacterial cell and reduce the ATP synthase activity thus decrease the metabolism processes of bacterial cell. They also decline the ribosomal activity of bacterial cell. Gold nanoparticles with small size and large surface area have more antibacterial activity as compared to large size particles (Ravishankar et al. 2011).

Bacterial cell wall and cytoplasm proteins are responsible for the functioning of the cell. Gold nanoparticles disturb the working of these proteins thus cause the death of the cell. The phosphorus containing DNA bases and sulphur containing proteins are the favorite sites of Gold nanoparticles in bacterial cells to react. They bind with the thiol group of enzymes in bacterial cells, disturb their respiratory chains and due to the high ROS production oxidative stress occurs. When gold nanoparticles are synthesized with the help of any capping agent then it increases their antimicrobial potential (Fig. 7.3).

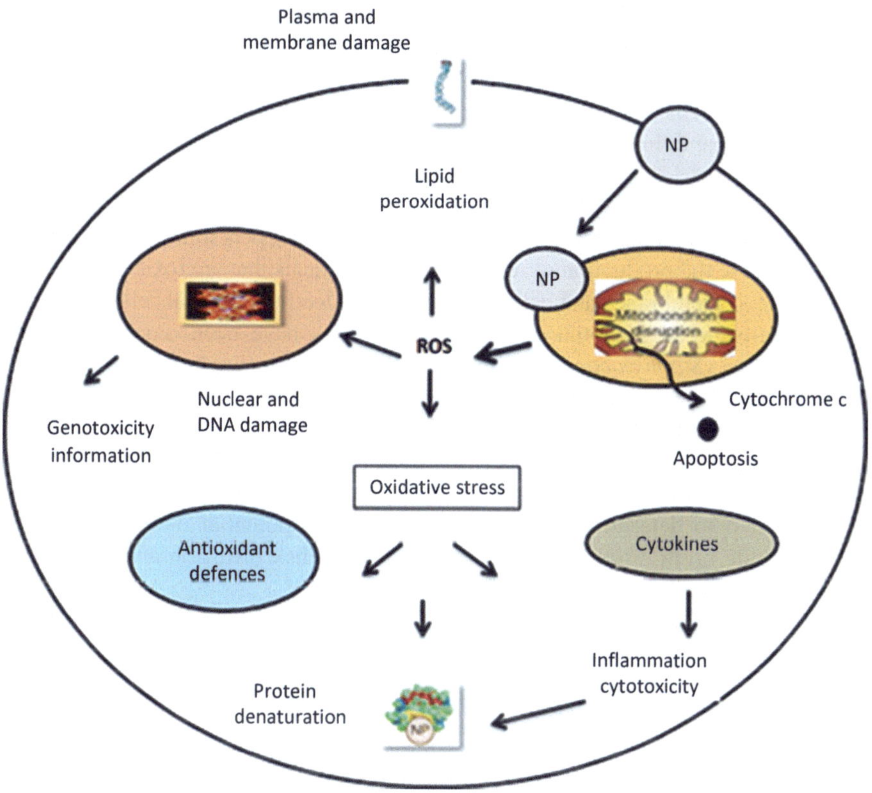

Fig. 7.3 Gold nanoparticles antimicrobial effect on bacterial cells

7.5 Effect of Metal Oxide Nanoparticles on Bacterial Cells?

Metal oxides nanoparticles such as zinc, iron, copper, titanium and magnesium oxide nanoparticles have been used in medical research due to their various medical properties. These particles have small size and good stability. The antimicrobial effect of these nanoparticles has been reported in Gram negative i.e., *Escherichia coli* and *Pseudomonas aeruginosa* and *Gram positive i.e., Staphylococcus aureus and Bacillus subtillis* bacterial strains. Zinc and copper oxide nanoparticles have high antimicrobial potential as compared to iron oxide nanoparticles.

7.5.1 Antibacterial Potential of Zinc Oxide Nanoparticles

Zinc oxide has been extensively used in creams and lotions due to their dermatological applications while nanoparticles of ZnO have much more antimicrobial potential as compared to Zinc oxide. The previous studies have shown that the small size Zinc oxide nanoparticles have great antibacterial potential as compared to large size particles. Several mechanisms have been reported on the antibacterial potential of zinc oxide nanoparticles. For example, they can interact with the membrane lipids of bacterial cells and disorganize the cell membrane structure with loss of membrane integrity and finally the bacterial cell death (Zhang et al. 2007).

ZnO nanoparticles are semiconductor in nature and can be activated under normal light or in small amount of UV-light. Thus they show good antibacterial effect in light as compared to dark conditions. ZnO nanoparticles also penetrate in the bacterial cells and cause the production of reactive oxygen radicals that are toxic and damage the DNA, cell membranes and cellular proteins that lead to bacterial cell death.

The antibacterial potential of ZnO nanoparticles have been reported in both Gram positive (*Staphylococcus aureus* and *Bacillus subtillis)* and Gram negative (*Escherichia coli*) bacterial cells. Zinc ions in zinc oxide nanoparticles interact with negatively charged cell wall of bacteria. It causes the damage of bacterial cell wall. These nanoparticles also interact with the DNA binding proteins and enzymes of bacterial cell that ultimately lead to bacterial cell death. Different methods have been reported to determine the *in vitro* antibacterial potential of bacterial cells. These include broth dilution method, agar dilution, disk diffusion and micro-titer plate based method.

7.5.2 Mechanism of Antibacterial Action of Zinc Oxide Nanoparticles

The photo-activation of ZnO nanoparticles is the major cause of their antibacterial activity. Under UV light exposure, the zinc nanoparticles increase the ROS production. These ROS cause the destruction of bacterial cell components such as proteins, lipids and DNA as a result of their internalization in bacterial cell. The ROS production has been reported in dark as well as in light condition in both Gram positive and Gram negative bacteria.

ZnO nanoparticles are bactericidal and by disrupting the bacterial cell membrane cause the membrane dysfunction and internalization of particles in bacteria. ZnO nanoparticles internalization depends upon particle size and surface chemistry. Nanoparticles form pits and holes in the bacterial cell wall that cause the cell wall disruption and extrusion of cytoplasmic contents. Positively charged zinc oxide nanoparticles attracts with the negatively charged bacterial cell wall, it causes the inter cellular reduction of components along with generation of ROS species and disturbance of DNA replication (Fig. 7.4).

The antibacterial mechanism of zinc oxide nanoparticles depend on multiple factors:

(a) UV illumination effect
(b) Impact of ZnO particles morphology
(c) Influence of ZnO nanoparticles size and concentration

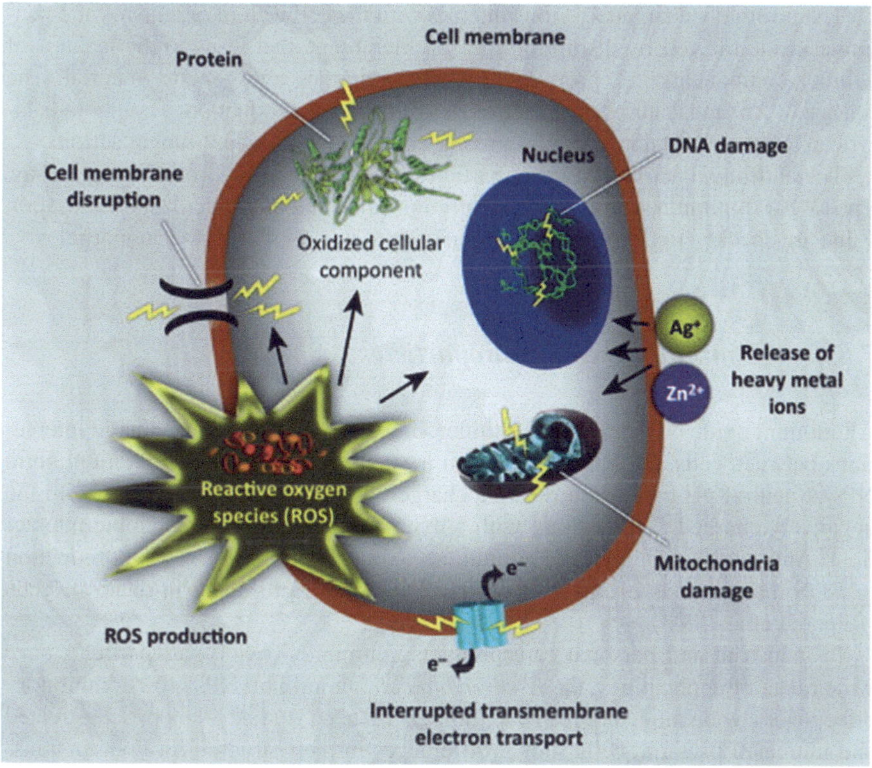

Fig. 7.4 Antibacterial effect of zinc oxide nanoparticles on bacterial cells

(d) Surface modification of nanoparticles
(e) Excessive ROS generation
(f) Electrostatic interaction between bacterial cells and NPs

All these factors influence the antibacterial potential of zinc oxide nanoparticles at different levels. Due to the strong potential of ZnO nanoparticles as antibacterial agent they are widely used in biomedical research applications and in food packaging industries. ZnO nanoparticles can also be used against multi-drug resistant bacteria and fungi hence can be used as a substitute to antibiotics.

7.5.3 Antifungal Properties of Zinc Oxide Nanoparticles

The antifungal activity of ZnO nanoparticles strongly depend upon the concentration and nanoparticles size. The small size of nanoparticles depicts good antimicrobial action due to its large surface to volume ratio.

The positively charged Zn^{+} ions interact with the cell wall of organisms and penetrate inside the cell by disrupting the cell membrane that leads to the leakage of cellular components i.e., proteins, enzymes, minerals and genetic material. The release of Zn^{+} ions causes the reactive oxygen species production. The photo catalytic activity of ZnO nanoparticles greatly contribute to their antifungal action.

The antifungal activity of these particles is determined by broth dilution assay. The (MIC) minimum inhibitory concentration value and microbicidal concentration value are checked to determine the antimicrobial potential of ZnO nanoparticles.

7.5.4 Aluminium Oxide Nanoparticles

Aluminium oxide NPs have high antimicrobial effect due to surface charge interactions between cells and NPs. Aluminium is stable at high temperature, aluminium NPs at neutral pH contains a positive charge on its surface that interacts with the negatively charged *E. coli* along with polymer bridging and hydrophobic interactions. Antimicrobial property of aluminium oxide NPs is because of the production of ROS that grounds disorder of cell wall in addition to result in death of cells (Rupareli et al. 2008).

In industrial and personal care products, aluminium oxide nanoparticles have wide range of applications. On *Escherichia coli*, aluminium NPs growth inhibitory effect over a wide range of (10–1000 μg/mL) has been studied. Nanoprticles treated and untreated bacterial cells shown differences in their structure in Fourier transform infrared studies.

At very high concentrations, aluminium nanoparticles exhibited a minor growth-inhibitory effect. It occurs due to the surface charge exchanges between the cells and particles. The free-radical scavenging properties of the particles might not be a reason of cell wall disruption (Sadiq et al. 2009).

Aluminium has corundum structure, oxygen atoms fits in hexagonal adjacent packing and in the lattice Al_{3+} fills two third of the octahedral sites. Also over a wide range of temperature it is thermodynamically stable (Florez et al. 2003). At near neutral pH, aluminium nanoparticles have a positive charge on their surface. Due to the electrostatic forces among the negatively charged *E. coli* and nanoparticles, adhesion occurs on the bacterial surfaces (Li and Logan 2004).

In the suspension solution concentration of particles increase that increases their adhesion on bacterial cells. With respect to concentration a negative effect on growth occur. The adhesion occurs due to the electrostatic interaction between the particles and bacterial surfaces, also the polymer bridging and hydrophobic interactions between them are responsible for it. The reactive oxygen species (ROS) generation is the reason of cell wall disruption and subsequently cell death and responsible for the antimicrobial property of these particles (Rupareli et al. 2008). As free radical scavengers, these NPs may act. Also from oxidative stress induced cell death these NPs rescue the cells.

7.5.5 Magnesium Oxide Nanoparticles

These nanoparticles aree highly ionic and can be synthesized with unusual crystal morphologies having different reactive sites and corners and high surface area (Stoimenov 2002). MgO nanoparticles have interesting property to absorb and hold significant amount of chlorine and bromine for a long time (Huang 2005). Against certain vegetative Gram negative and positive bacteria and spores, MgO/X2 nanoparticles showed biocidal activity (Richards et al. 2000). As a potent disinfectant, MgO nanoparticles possess many properties (Koper et al. 2002).

These nanocrystals can absorb active halogens due to their large surface area and increases surface activity. These particles can cover the bacterial cells to a large extent due to their small size and high concentration (Richards et al. 2000).

They have shown unique bactericidal activity against *Bacillus megaterium, E. coli* and *Bacillus subtilis* spores. As compared to the bacteria and spore cells, the MgO nanoparticles have high bioactivity in water suspension. The opposite charge brings the nanoparticles and bacteria composed as aggregates in water suspensions as shown via confocal microscopy studies. Electron microscopy and atomic force microscopy studies reveal that MgO nanoparticles treated with halogen i.e., bromine and chlorine have a rapid and strong effect on bacteria and spores killing (Koper et al. 2002).

7.5.6 Copper Oxide Nanoparticles

Copper oxide has a monoclinic structure and is a semiconducting compound. It is the simplest copper compound and has many valuable physical properties such as superconductivity, high temperature and spin dynamics. Copper oxide in crystal form also has photovoltaic, photo catalytic properties and photoconductive functionalities (Kwak and Kim 2005). The antimicrobial mechanism of nanoparticles of CuO is still not clear. CuO has wide applications as it is economical than silver, physically and chemically stable and easily mixes with polymers (Xu et al. 2009). Copper oxide nanoparticles are highly ionic and they can be prepared with unusual crystal morphologies with high surface areas. Therefore they can be used as antimicrobial agents (Stoimenov 2002).

Nano copper oxide is required in high concentration for their bactericidal effect. These nanoparticles proved to be active for a variety of bacterial pathogens killing that cause hospital related infections (Ren et al. 2009). Gram negative bacteria Pseudomonas *aeruginosa* and *Proteus* specie are less prone to positively charged antimicrobials due to less peptidoglycan in their cell wall.

Various studied have been performed to determine the nanoCuO potential sur rounded by different polymer materials. High release killing ability was observed against MRSA strains, as compared with lower contact killing ability. For the optimal antimicrobial activity, the discharge of ions into the local surroundings is necessary (Cioffi et al. 2005).

Against *Bacillus subtilis*, copper oxide nanoparticles have large antimicrobial activity. This may occur due to the high amount of carboxyl and amines groups present on the *B. subtilis* surface and copper has high attraction to these groups. The released copper ions may also bind with nucleic acid stand and DNA molecules. Inside the bacterial cells these copper ions disrupt the biochemical processes (Rupareli et al. 2008). The particular bactericidal effect of copper nanoparticles is not clear.

7.5.7 Titanium Dioxide Nanoparticles

When TiO_2 nanoparticles are illuminated at wavelength of less than 385 nm in UV light, their inhibitory activity occurs due to high photo catalytic oxidation (Chorianopoulus et al. 2011).

On illumination near the UV light, TiO_2 particles induce the death of bacteria. These photoexcited TiO_2 nanoparticles generates active free hydroxyl radicals (OH^-) that possibly responsible for their antibacterial activity (Wei et al. 2004). In water, against Escherichia coli and other bacteria and fungi, the antimicrobial effect of TiO_2 photocatalyst has been determined (Chawengkijwanich and Hayata 2008).

The nitrogen-doped metal oxide nanocatalysts bactericidal activity has also been reported on *E. coli* biofilms and on components of TiO_2 coated biofilm surfaces. Hence it is determined that TiO_2 photo catalysts used as alternate way of self-cleaning of the filthy surfaces and more improvement may offer strong purifying solutions for inhibition of biofilms. In food processing industries, TiO_2 photocatalysts can be used as operative biofilm sterilizer (Liu et al. 2007). For *E. coli* killing, suspensions containing TiO_2 are effective. This has led to use TiO_2 in aqueous media for the killing of bacteria and viruses by developing photocatalytic methods (Duffy et al. 2002).

It is suggested that under UV radiation the nanostructured TiO_2 can be used actively and it eliminates pathogenic microorganisms in food contact surfaces and reduce disinfection time and increase food safety (Chorianopoulus et al. 2011). The disadvantage of using TiO_2 is that to activate the photocatalyst, the UV light is required that can kill the bacteria and viruses (Hu et al. 2006; Sunada et al. 1998) (Table 7.1).

7.5.8 Antiviral Impact of Nanoparticles

The antibacterial potential of metal and metal oxide nanoparticles has been widely studied but their anti-viral properties are not fully discovered (Sun et al. 2005). Viruses cause diseases that affect the whole world socially and economically. To

Table 7.1 Antimicrobial nanoparticles of different nanoparticles

Properties	Mechanism of action	Examples of nanoparticles
Antibacterial	Interaction with phosphorus moieties in DNA, resulting in inactivation of DNA replication.	Silver nanoparticles have inhibitory activity against *E. coli, Bacillus subtilis, Staphylococcus aureus*, methicillin-resistant coagulase-negative staphylococci, vancomycin-resistant *Enterococcus faecium*, ESBL-positive *Klebsiella pneumonia, S. typhi, Vibrio cholera.*
	Reacts with the silver containing proteins, leading to the inhibition of enzymes function.	Gold nanoparticles have antibacterial activity against MRSA, VRE, *E. coli, Pseudomonas aeruginosa.*
		MgO nanoparticles have excellent against *E. coli, B. subtilis, B. megaterium.*
		CuO strongly inhibits *B. subtilis.*
		Aluminium oxide nanoparticles have growth inhibitory effect on *E. coli.*
		TiO2 nanoparticls are effective in killing *E. coli, S. aureus, Listeria Monocytogenes.*
		ZnO nanoparticles inhibit food-borne bacteria *E. coli* 0157:H7, *B. subtilis, Pseudomonas fluorescens, L. monocytogenes, Salmonella enteritidis, S. aureus, S. typhimurium.*
Antiviral	Blocking of viral attatchment to the cell surface.	Gold nanoparticles have anti-HIV activity and inhibit several strains of influenza virus.
		Silver nanoparticles inhibit HIV-1, influenza virus, Herpes Simplex virus, Respiratory syncytial virus, Monkey pox virus.
Antifungal	Disruption of cell membrane.	Silver nanoparticles have fungicidal and fungistatic effects on the dermatophytes *Trichophyton mentagrophytes* and *Candida species.*

develop the drugs that can target the virus and to maintain the host cell sustainability is a big challenge (Pinto et al. 2009).

Due to their core material and ligand shells, metal nanoparticles have been recommended as anti-viral systems. At non-cytotoxic concentrations, anti-viral activity of silver nanoparticles proved to work against HIV (Lara et al. 2010).

The antiviral approach of AgNPs against HIV was explained by using a number of *in vitro* assays. It was discovered that as an inhibitor of viral entry or as virucidal agent, silver nanoparticles exert anti-viral activity, at an initial stage.

For their anti viral and anti-HIV activity, gold nanoparticles have also been studied. Production of multiple ligands using nanoparticles produces binding molecules in high concentration that can help in biological interaction targeting. As alternate HIV systems, known for the targeting of different stages of virus infection, multivalent gold nanoparticles coated with sulfate ligands are recommended (Giancivincenzo et al. 2010).

Similarly on the nano gold platform, sulfate ligands and other active particles that target the HIV life cycle at different stages simultaneously modified which makes gold nanoparticles as attracting support and it helps in the formation of new multifunctional anti-HIV systems (Bowman et al. 2008). To inhibit the influenza, gold nanoparticles with different anionic groups have been used. Against different influenza strains, these nanoparticles have exhibited effective influenza properties.

Because of the blockage of virus assembly on the cell surface, main mechanism of inhibition occurs. The charge density and the functional groups of the nanoparticles have an essential role in inhibition due to the anionic groups. On the surface of anionic gold functionalized groups end groups present that can join with the viruses with their multivalent bonds. In cell, gold nanoparticles can enter via endosome vesicle due to their small size and they can also delay the fusion step. Several studies revealed that gold nanoparticles have low or no cytotoxicity and if they are cytotoxic, toxicity depends on size, shape and particles surface charge (Simpson et al. 2010).

The nanoparticles formed with the gold core and via thiol group mercapto ethane sulfonate (MES) molecules attached with its surface shown no cytotoxicity and it can inhibit several influenza viruses (including the pandemic swine influenza A (H1N1) strain) which indicate that its contact is not restricted to a single strain. The inhibition effect of silver nanoparticles capped with mercaotorthanesulfonate has been studied against Type 1 infection of Herpes Simplex Virus (Pinto et al. 2009).

Against the Hepatitis B, HIV 1, respirational syncytial and money pox virus, the antiviral action of silver nanoparticles has been stated (Lu et al. 2008). All these studied proposed that metal nanoparticles with various active groups have better antiviral action and they aid in fighting viral diseases.

7.5.9 Applications of Nanoparticles

Against common pathogenic microorganisms, metal based nanoparticles proved to be an effective antimicrobial agent. Also as antimicrobials and spices in food, industrial products, customer and medicinal products zinc oxide, silver and titanium oxide are receiving considerable attention (Dibrov et al. 2012). Silver nanoparticles are widely combined with several devices and materials to enhance their value and biomedical applications due to their broad spectrum antimicrobial potential (Sarkar et al. 2007).

In health linked products such as catheters, dressings and other material especially during the wound healing process and burns, nanoparticles are used as additives to stop infection. To increase their antimicrobial properties, nanoparticles are added in much commonly used stuff such as in tooth paste, water purifiers, fabrics, shampoo, paints, deodrants, filters, toys, kitchen utensils and humidifiers (Baker et al. 2005).

Titanium oxide nanoparticles are utilized in cosmetics, in filters to remove orders and exhibit their germicidal properties and as antimicrobial agent in combination with silver. They are also used in waste water treatment due to their photo catalytic properties. They are allowed by the American Food and Drug Administration (FDA) to utilize in human cosmetics, medicines, foodstuff and ingredients as they are nontoxic (Wist et al. 2004).

Nowadays, as a self-purifying material for surface coating titanium oxide nanoparticles have wide applications and utilized as antiseptic tools for food industries. Due to their antimicrobial properties, zinc oxide and copper oxide nanoparticles are being fused with different skin and medical coatings. As antimicrobials, they are used in hospital wallpapers. With reference to its antibacterial properties, ZnO powder is used as an active component in balms, ointments and lotions for their dermatological applications (Flores et al. 2003).

7.6 Drug Delivery and Nanoparticles

7.6.1 Introduction

In the recent area, exceptional growth has been witnessed in the area of Nano science and nanotechnology. It is considered that as practical to medicine, nanotechnology will bring major advances in the analysis and cure of disease. There is increasing optimism that nanotechnology, as applied to medicine, will get significant advances in the diagnosis and treatment of disease (De Jong et al. 2005; Ferrari 2005; Duncan 2003).

To understand these applications, engineered nanoparticles are significant tool. For use in medical applications, all the particles do not fulfill the recommended standards and now their mostly acknowledged definition is "the particles in the size of 100 nm (The Royal Society and Royal Academy of Engineering 2004).

Because of their distinctive and significant functions i.e., their quantum properties, their large surface to mass ration than other particles and their ability to carry and absorb other compounds, these nanoparticles are beneficial for healing purposes. Relatively NPs have large functional surface area which can adsorb, bind and transfer other composites such as probes, pills and proteins. However for the practical applications of nanotechnology many challenges must be overcome a) to understand the pathophysiological base of infection, get advanced diagnostics prospects and improves therapies for yield. Although nanoparticles defined as particles having dimensions below 0.1 μm or 100 nm but relatively large size particles (size more than 100 nm) are needed in drug delivery. Also, the drug may be manufactured at Nano scale and work as its own carrier along with the engineered particles (Kipp 2004; Duncan 2003; Baran et al. 2002; Cascone et al. 2002).

The structure of the nanomaterial's may differ according to the source material used i.e., of biological origin like lipids, chitosan, dextran, phospholipids, lactic acids or have other chemicals like metals, silica, carbon and various polymers. The organic compounds like phospholipids contact with cells will be changed as paralleled to non-biological constituents such as cadmium and iron. There is an extensive area of possibilities of polymer engineered nanoparticles due to their chemical composition.

For the drug targeting solid nanoparticles may be used, the drug carrier must be released when they reach the targeted spot in the body. For the active movement and discharge of the drug in drug delivery, ecological nanoparticle formulations are necessary. Data has been collected from the studies on inhalation of nanoparticles to understand the toxicity and biological behavior of particles. The inhalation of particles may occur due to the diesel exhaust particles or unplanned discharge of ultrafine particles (Oberdörster et al. 2005; Dreher 2004; Kreyling et al. 2004; Donaldson and Stone 2003; Borm 2002; Donaldson et al. 2001; Obordoster 1996).

Investigation has proved that contact with these ultrafine particles is connected with range of impacts (Donaldson et al. 2005) comprising immune adjuvant effects, pulmonary inflammation (Granum and Lovik 2002) and cardiovascular and blood coagulation (Oberdörster et al. 2005; Borm and Kreyling 2004). For both nanoparticles and ultrafine particles the cut off size is 100 nm so both are considered as equivalent. The ultrafine particles are part of environmental effluence so it is supposed that engineered nanoparticles may have similar effects.

7.6.2 Nanoparticles and Drug Delivery

In the Nanomedicine background, pharmaceutical and drug delivery development must be considered as the science and technology in nanometer scales (10–1000 nm) (Ferrari 2005; Duncan 2003).

The nanoparticle drug preparations are also possible (Kipp 2004; Duncan 2003; Baran et al. 2002; Cascone et al. 2002). This also indicates the formulations of smart drugs or the agnostics by diagnosing, preventing and treating diseases (Lavan et al. 2003). In drug delivery, the major areas of nano-biotechnologies contain:

- Specific drug delivery and targeting,
- Decrease of toxicity while retaining the healing effects,
- Provide more protection and biocompatibility, and
- Fast development of new and harmless medicines.

There are some fundamentals problems for the designing of new materials that can be used as carriers for drug delivery systems. You must have knowledge on (a) drug absorption and release (b) maintains system constancy and lifecycle (c) bio dispersal (d) biocompatibility (e) functionality. Also the possible harmful effects of the remaining substance that is used as carrier should be kept in mind. In this opinion, therapeutically needed biodegradable nanoparticles with short life span would be ideal.

As drug carriers most used composites are eco-friendly polymers that releases drug after degradation. In liver and spleen of humans, a setup is present that causes the difficulties in the use of nanomaterial as drug carriers. But while for the treatment of liver diseases like hepatitis and tumor metastasis, targeting of liver by nanoparticles may be advantageous (Moghimi et al. 2001).

Nanoparticle frame of drugs aims at (1) increase drug delivery and uptake (2) target specific cells (3) Decrease in the drug and non-targeted organs toxicity. These will raise the therapeutic index and discovery of drug doses that result in tumor cell death without harmfulness to other structures. To achieve this, formation of target specific and nanoparticles with long life-span are required. Polyethylene glycol

(PEG) is used for nanoparticles surface modification that caused their sustained presence in blood flow and inhibition of phagocytosis by the mononuclear phagocytic system (Peracchia et al. 1999; Bazile et al. 1995).

An *in vitro* study revealed when gold nanorods were transformed with PEG, their toxicity also decreased (Niidome et al. 2006). To inhibit the cluster formation of NPS coating may also require. Several materials can be used for coatings to keep the particles in colloidal form and to stop the cluster formation. As coatings several synthetic polymers i.e., poly vinylpyrroldine (PVP), polyethylene glycol (PEG), various natural polymers like chitosan, dextran and pullutan etc. and surfactants like dodecylamind and sodium oleate are used.

Nanoparticles distribution can be effected by NPs size as it was verified for lipid v esicles. The smaller vesicles (200/300 nm versus 25/50 nm) decrease the liver uptake (Seki et al. 2004). Also the minor differences in size can affect their bioavailablity and actual distribution (Fang et al. 2006; Shim et al. 2004; Zhang et al. 2004; Fishbein et al. 2001; Saez et al. 2000).

The approval rate by mononuclear phagocytic system increases for liposomes of 100 nm in size while charges were more important for liposomes below 100 nm in size (Senior et al. 1985; Senior and Gregoriadis 1982). The composition will be important for all the particles below 100 nm in size. The actual shape and size of nanomaterials will be important apart from the earlier findings on their form.

7.6.3 Nanoparticles Use in Drug Delivery

The use of nanoparticles at the targeted size for drug delivery without any size effects is the major challenge. During cancer, tumor spreads in different organs and it is difficult to treat it by using drugs. Chemotherapeutics cannot be used at their full therapeutic potential due to their cytotoxic potential. Increased levels of drugs are required for local or targeted drug delivery. Specific nanoparticles are used as tools for this purpose. Small sizes of nanoparticles give them advantage to pass through the cell membranes.

The use of nanosized chemotherapeutics formulations has been investigated since a long time. The nanosized phospholipid fatty structures (liposomes) are able to pass along smallest arteriols without clotting due to their small size, flexibility and compatibility. The distribution of nanocapsulated drugs can be altered by using different polymers and dendrimers.

In particle uptake, effects and distribution, surface chemistry of particles is important. Quartz was surface modified with PVNO-polymer and when uptake via macrophages, no cytoxicity and genotoxicity shown in epithelial cells during acute and chronic inflammation. When polystyrene particles were surface modified they have various effects on coagulation of blood, cellular oxidative burst and on ROS formation in mitochondria (Xia et al. 2006). The circulation time of nanoparticles can be increased by surface coating with polyethylene glycol (PEG) (Nidome et al. 2006).

7.6.4 Drug Delivery of Nanoparticles Using Cellular and Intracellular Targets

Drug delivery transforms the drug release profile, its concentration, dispersal and exclusion to enhance its efficiency. In the range of drug delivery, current focus is on the improvement of targeted delivery i.e., in the body only in the target area the drug is active and persistent release of drug formulations in which drug is released in the body in controlled manner at specific intervals.

For medical use, numerous therapeutic nanocarriers have been accepted. Over drugs, these nanocarriers have many advantages. They

- Defend the drug from early degradation
- Inhibit the too early contact of drug with the biological environment
- Increase the drug absorption in a specific tissue (e.g, in tumor).
- Control the drug delivery in tissues and pharmacokinetic.
- Enhance drug intracellular dispersion.

Until now, there are some therapeutically accepted nanocarriers that comprise the molecules that bind selectively and target cancer cells. Presently due to their distinctive physiochemical properties metallic nanoparticles gain a lot of attention. Silver has the maximum thermal and electrical conductivity amongst all metals. For specific targeting of cancer cells, nano silver act as a carrier. In the normal cells usually the pores are less than 10 nm in size and in the cancer cells the pores are 10–100 nm (Soumya and Hela 2013).

Surface modification and functionalization of gold nanoparticles occur by using PEG (polyethylene glycol), it increases the efficiency of gold nanoparticles and contribute in its localization in nuclear region. Quantum dots are coated with polymer shells that help them in targeting different molecules. For vasculature of lungs and tumor, very small size quantum dots coated with peptide coated dots have been used for specific targeting of cells (Akerman et al. 2002).

Surface modified nanoparticles are necessary specifically to use them in medicinal uses like in cellular uptake, cellular binding, drug targeting and intracellular support. These modifications increase the activity of drugs loaded on the nanoparticles. Adherence of specific proteins on nanoparticles surface increase the activity of these nanoparticles and aid in immunologically direct targeting of the nanoparticles (Nobs et al. 2004).

7.6.5 Toxicological Possibilities of Nanoparticles

The knowledge of the nanoparticles/ nanotechnology safety and toxicological issues is very important before their use in Nanomedicine. In pharmaceuticals industry, to raise their therapeutic ratio particular drug delivery formulations are used by measuring their efficacy and toxicity. However for all these formulations including nanoparticles their toxicological evaluation is necessary before their use in medical applications. Different drug formulations are harmful as compared to nanoparticles and it is proven in different studies (Ferrari 2005).

As compared to bulk materials nanoparticles have specific properties. The surface of nanoparticles comes in contact with the body tissue during specific targeting so the properties of these nanoparticles need to be investigated. The properties of nanoparticles depend upon a number of factors like the origin of these particles i.e., biological and non-biological.

Nanoparticles have qualitatively different physiochemical characteristics as compared to micron-sized particles therefore they may functions differently in the body. Toxicity levels of these nanoparticles are investigated by using different assays. Nanoparticles that are used as drug carriers can reduce the drugs toxicity incorporated in nanoparticles. The entire formulation toxicity is examined. There is a need of new assays and techniques to investigate these nanoparticles (Scenihr 2006).

Nanoparticles have large surface area to volume ratio owing to their small size. Therefore they have specific effects on cells, blood, cardiovascular system and other organs of the body. Engineered ligand coated nanoparticles are used in drug delivery as agents. Some particles enter in tissue without affecting its function while others do not do that. Positively charged gold and polystyrene particles have been exposed for blood clotting and hemolysis while negatively charged nanoparticles are non-toxic.

7.7 Newly Engineered nanomaterial's Toxicology Level

Newly engineered nanomaterials include carbon nanotubes, quantum dots, fullerenes, dendrimrs, gold nanoparticles, nanoshells and silica nanoparticles. Apart from size and surface area, surface properties of nanoparticles have been studied by different researchers.

7.7.1 Carbon Nanotubes

Carbon nanotubes are long tubes made up of single or multiple grapheme layers and are called single or multiple walled carbon nanotubes. These tubes have potential to act as bio persistent fiber. Both single walled or multiple walled carbon nanotubes vary in length from several mm and they differ in diameter.

During *in vitro* growth of bronchial epithelium cells, high quantities of single walled carbon nanotubes influences ROS production, rate of lipid peroxidation and oxidative stress and modifications in cell characteristics (Sayes et al. 2006).

Latest studies reported that both single and multiple walled carbon nanotubes cause platelets aggregation but not for the building blocks of carbon nanotubes i.e., C60 fullerenes Multiple walled carbon nanotubes (MWCNT) also exhibit pre-inflammatory effect on keratinocytes (Monteiro-Riviere et al. 2005). These carbon nanotubes are more toxic on dose per mass basis as compare to quartz particles responsible for lung toxicity.

7.7.2 Fullerenes

Fullerenes have been studied as potential antimicrobial agents as they cause the production of reactive oxygen species after photoexicitation (Yamakoshi et al. 2003). Fullerenes are the important building blocks of nanomaterials and they are usually made up of $C_{60.}$ They also affect the brain cells by causing lipid peroxidation and generation of ROS by microglia. Cytochrome P450 breakdown affect the brain cells by the production of reactive fullerenes metabolites.

7.7.3 Dendrimers

Due to their particular nature, dendrimers are suitable for drug delivery. Because of their small size (upto 10 nm) they have less drug loading capacity but their branching and dendritic nature allow the loading of drug on their polymeric structures (Svenson and Tomalia 2005). Their surface modification and functionalization using antibodies may enhance their specific targeting potential. Dendrimers have also been studied as antimicrobial agents, trans-membrane transport and also as gene transfer agents. Dendrimers are only safe when use for specific medical application otherwise they are toxic.

7.7.4 Quantum Dots

Quantum dots are diverse cluster of nanoparticles (Hardman 2006). They may differ in size, ranges from 2.5 upto 100 nm. Multiple factors affect the quantum dots absorption, distribution, excretion, metabolism and toxicity. Inherent physiochemical properties and environmental conditions also affect these properties of quantum dots. Quantum dots size depend upon their coating thickness. Only few studies have been performed to investigate the toxicity of quantum dots.

Due to their apparent coating quantum dots verified as toxic in *in vitro* studies (Hoshino et al. 2004). Surface modification of quantum dots with N-acetylcysteine reduced their toxicity while non-modified quantum dots made up of cadmium telluride cause lipid peroxidation in the cells (Choi et al. 2007). Quantum dots without any surface modification can cause the production of reactive oxygen species and injury of plasma membrane, mitochondria and nucleus. The bioactive covering used for surface modification is necessary for exact targeting of cells and organelles.

The core material of quantum dots is also responsible for its toxicity as cadmium containing quantum dots are highly toxic due to the free, highly toxic cadmium ions. Quantum dots lack both *in vitro* and *in vivo* toxicity as reported by Hardman 2006. However more investigation is required to study the toxicological effects of quantum dots.

7.7.5 Gold Nanoparticles/Nanoshells

The colloidal, metallic gold nanoparticles are extensively used. They can be manufactured in various forms (dots, rods) and are detected at very low concentrations. These are commercially available in various sizes. Gold nanoparticles have no cytotoxic effects on cells (Cornor et al. 2005). They are widely used in biomedical applications: as possible carriers in drug delivery, imaging of genes and molecules (Kawano et al. 2006). They are also used for the generation of new cancer healing products (Hirsch et al. 2003; Loo et al. 2004).

The stabilizer CTAB is responsible for the cytotoxicity of gold nanorods. After removing the excess of CTAB, the PEG modified gold nanorods does not show any cytotoxicity (Niidome et al. 2006). Nanogold suspension was evaluated for an acute oral toxicity study and no signs of injurious effects were noted (Lai et al. 2006).

Nanoshells are composed of gold-copper or gold-silver and for this purpose gold solutions are used. As contrast agents, gold-copper, silver-gold are used in magnetic resonance imaging (RMI). Gold-silica formulations are used for photo-thermal removal of tumor cells. In tumor cells during *invitro* studies, non-targeted nanoshellsdo not exhibit any cytotoxicity. After laser activation, these nanoshells bind to the tumor cells and cells death occurs (Stern et al. 2007).

7.7.6 Silica Nanoparticles

In *in vitro* studies both toxic and non-toxic properties of silica nanoparticles are observed. The dose dependent cytotoxicity of different size silica nanoparticles were obtained *in vitro* (Lin et al. 2006). The toxicity is dependent on dose and exposure time (24, 48, 72 h) of silica nanoparticles. By increasing dose and exposure time, the toxicity of silica nanoparticles also increases. SiO_2 nanoparticles exposure to cells resulted in increased oxidative stress by increasing ROS level and by reducing glutathione levels.

Silica nanoparticles at high dosage cause decrease in cell viability and release of enzyme lactate dehydrogenase from the cells that causes the membrane damage. Therefore silica nanoparticles prove to be toxic for the cells (Chang et al. 2007). In concentration above 0.1 mg/ml, major decrease in cell viability was observed. After surface modification of silica nanoparticles, no or less cytotoxicity was noticed (Ravi Kumar et al. 2004).

7.8 Needs of Nanomaterials in Medicine

Several studies have been reported the cytotoxicity of nanoparticles but they are mostly based on small panel of nanoparticles. One general conclusion is that the small size nanoparticles are more toxic as compared to large size nanoparticles with same composition. In various studies, nanoformulation drugs are studied with their possible toxicity effects, only drug toxicity level is considered and possible toxicity of drug carriers (i.e, nanoparticles) is not studied.

To analyze the potential hazards of nanomaterials, certain assays need to be performed. It will help in their possible biomedical applications. In 2006 Costigan reviewed the toxicity of nanoparticles in healthcare products. Each nanoparticle formulation must be tested to investigate its toxic potential. Different assays and techniques are used dependent on case basis.

For the development of testing protocols and procedures, various issues need to be considered. Specific effects of nanoparticles, their mechanism of action, risks of using these nanomaterials, their physio-chemical and other properties are considered. For decades, nanoparticles have been used to reduce the toxic potential of drugs and their side effects. Nanomaterials act as drug carriers and until now, no side effects of these nanomaterials have been reported. There is a need to investigate this issue in detail for better understanding of the possible toxicity mechanism of nanomaterials used as drug carriers.

References

Ahmed S, Ahmad M, Swami BL, Ikram S (2016) Plants extract mediated synthesis of silver nanoparticles for antimicrobial applications: a green expertise. J Adv Res 7:17–28

Åkerman ME, Chan WCW, Laakkonen P (2002) Nanocrystal targeting in vivo. PNAS 99:12617–12621

Ankana S, Prasad TNVKV, Elumalai EK, Savithramma N (2010) Production of biogenic silver nanoparticles using *Boswellia ovalifoliolata* stem bark. Dig J Nanometer Biostruct 5:369–372

Baker C, Pradhan A, Pakstis L, Pochan DJ, Shah SI (2005) Synthesis and antibacterial properties of silver nanoparticles. J Nanosci Nanotechnol 5:244–249

Balouiri M, Sadiki M, Ibnsouda SK (2016) Methods for *in vitro* evaluating antimicrobial activity: a review. J Pharm Anal 1:1–9

Baran ET, Ozer N, Hasirci V (2002) *In vitro* half-life of nanoencapsulated L-asparaginase. J Mat Sci Mat in Med 13:1113–1121

Bazile D, Prud'Homme C, Bassoullet MT (1995) Stealth PEG-PLA nanoparticles avoid uptake by the mononuclear phagocytes system. J Pharm Sci 84:493–498

Bhattacharya R, Mukherjee P (2008) Biological properties of "naked" metal nanoparticles. Adv Drug Del Rev 60:1289–1306

Borm PJ (2002) Particle toxicology: from coal mining to nanotechnology. Inhal Toxicol 14:311–324

Borm PJ, Kreyling W (2004) Toxicological hazards of inhaled nanoparticles-potential implications for drug delivery. J Nanosci Nanotechnol 4:521–531

Bowman MC, Ballard TE, Ackerson CJ, Feldheim DJ, Margolis DM, Melander C (2008) Inhibition of HIV fusion with multivalent gold nanoparticles. J Am Chem Soc 130:6896–6897

Cascone MG, Lazzeri L, Carmignani C (2002) Gelatin nanoparticles produced by a simple W/O emulsion as drug delivery system for methotrexate. J Mat Sci Mat in Med 13:523–526
Chang JS, Chang KLB, Hwang DF (2007) *In vitro* cytotoxicity of silica nanoparticles at high concentrations strongly depends on the metabolic activity type of the cell line. Environ Sci Technol 41:2064–2068
Chawengkijwanich C, Hayata Y (2008) Development of TiO_2 powder-coated food packaging film and its ability to inactivate *Escherichia coli in vitro* and in actual tests. Int J Food Microbiol 123:288–292
Choi AO, Cho SJ, Desbarats J (2007) Quantum dot-induced cell death involves fats up regulation and lipid peroxidation in human neuroblastoma cells. J Nanobiotech 12:1
Chorianopoulos NG, Tsoukleris DS, Panagou EZ, Falaras P, Nychas GJE (2011) Use of titanium dioxide (TiO_2) photocatalysts as alternative means for *Listeria monocytogenes* biofilm disinfection in food processing. Food Microbiol 28:164–170
Cioffi N, Torsi L, Ditaranto N, Tantillo G, Ghibelli L, Sabbatini L (2005) Copper nanoparticle/polymer composites with antifungal and bacteriostatic properties. Chem Mater 17:5255–5262
CLSI, Methods for dilution antimicrobial susceptibility tests for bacteria that grow aerobically, Approved Standard, 9th ed., CLSI document M07-A9. Clinical and Laboratory Standards Institute, 950 West Valley Road, Suite 2500, Wayne, Pennsylvania 19087, USA, 2012
Cornor EE, Mwamuka J, Gole A (2005) Gold nanoparticles are taken by human cells but do not cause acute toxicity. Small 1:325–327
Costigan S (2006) The toxicology of nanoparticles used in health care products. Available at the website of the medicines and healthcare products regulatory agency, Department of Health, UK. Accessed 20 November 2006. http://www.mhra.gov.uk/home/idcplg?IdcService=SS_GET_PAGE&nodeId=996
De Jong WH, Geertsma RE, Roszek B (2005) Nanotechnology in medical applications. Possible risks for human health. Report 265001002/2005, National Institute of Public Health and Environment (RIVM), Bilthoven, The Netherlands
Dibrov P, Dzioba J, Gosink KK, Hase CC (2012) Chemiosmotic mechanism of antimicrobial activity of Ag(+) in *Vibrio cholerae*. Antimicrob Agents Chemother 46:2668–2670
Donaldson K, Stone V (2003) Current hypotheses on the mechanism of toxicity of ultrafine particles. Ann 1st Super Sanita 39:405–410
Donaldson K, Stone V, Clouter A (2001) Ultrafine Particles. Occup Env Med 58:211–216
Donaldson K, Tran L, Jimenez LA (2005) Combustion-derived nanoparticles: a review of their toxicology following inhalation exposure. Part Fibre Toxicol 2:10
Dreher KL (2004) Toxicological highlight. Health and environmental impact of nanotechnology: toxicological assessment of manufactured nanoparticles. Toxicol Sci 77:3–5
Duffy EF, Touati FA, Kehoe SC, McLoughlin OA, Gill LW, Gernjak W, Oller I, Maldonado MI, Malato S, Cassidy J, Reed RH, McGuigan KG (2002) A novel TiO2-assisted solar photocatalytic batch-process disinfection reactor for the treatment of biological and chemical contaminants in domestic drinking water in developing countries. Sol Energy 77:649–655
Duncan R (2003) The dawning era of polymer therapeutics. Nat Rev Drug Disc 2:347–360
Fang C, Shi B, Pei YY (2006) *In vivo* tumor targeting of tumor necrosis factor-alpha-loaded stealth nanoparticles: effect of MePEG molecular weight and particle size. Eur J Pharm Sci 27:27–36
Ferrari M (2005) Cancer nanotechnology: opportunity and challenges. Nat Rev Cancer 5:161–171
Feynman R (1959) There's plenty of room at the bottom. A talk given at the annual meeting of the American Physical Society at the California Institute of Technology, December 29, 1959. A full transcript of the lecture can be found at www.zyvex.com/nanotech/feynman.html
Fishbein I, Chorny M, Banai S (2001) Formulation and delivery mode affect disposition and activity of tyrphostin-loaded nanoparticles in the rat carotid model. Arterioscler Thromb Vasc Biol 21(9):1434
Flores ME, Negrete J, Torres VG (2003) Structure and properties of Zn-Al-Cu alloy reinforced with alumina particles. Mater Des 24:281–286

Furno F, Morley KS, Wong B, Sharp BL, Arnold PL, Howdle SM (2004) Silver nanoparticles and polymeric medical devices: a new approach to prevention of infection? J Antimicrob Chemother 54(6):1019–1024

Giancivincenzo PD, Marradi M, Martinez-Avila OM, Bedoya LM, Alcami J, Penades S (2010) Gold nanoparticles capped with sulfate-ended ligands as anti-HIV agents. Bioorg Med Chem Lett 20:2718–2721

Granum B, Lovik M (2002) The effect of nanoparticles on immune allergic responses. Toxicol Sci 65:7–17

Gurunathan S (2009) Biosynthesis, purification and characterization of silver nanoparticles using *Escherichia coli*. Colloids Surf B Biointerfaces 74:328–335

Hardman R (2006) A toxicological review of quantum dots: toxicity depends on physiochemical and environmental factors. Environ Healh Perspect 114:165–172

Higby GJ (1990) Gold in medicine: a review of its use in the west before 1900. Gold Bull 15:130–140

Hirsch LR, Stafford RJ, Bankson JA (2003) Nanoshell-mediated near infrared thermal therapy of tumors under magnetic resonance guidance. Proc Nat Acad Sci USA 100:13549–13554

Hood JR, Wilkinson JM, Cavanagh HMA (2003) Evaluation of common antibacterial screening methods utilized in essential oils research. J Essent Oil Res 15(6):428–433

Hoshino A, Fujioka K, Oku T (2004) Physicochemical properties and cellular toxicity of nanocrystal quantum dots depend on their surface modification. Nano Lett 4:2163–2169

Hu C, Lan Y, Qu J, Hu X, Wang A (2006) Ag/AgBr/TiO_2 visible light photocatalyst for destruction of azodyes and bacteria. J Phys Chem B 110:4066–4072

Huang L (2005) Controllable preparation of Nano-MgO and investigation of its bactericidal properties. J Inorg Biochem 99:986–993

Kalishwaralal K, Banumathi E, Pandian SBRK, Deepak V, Muniyandi J, Eom SH (2009) Silver nanoparticles inhibit VEGF induced cell proliferation and migration in bovine retinal endothelial cells. Colloids Surf B Biointerf 73:51–57

Kawano T, Yamagata M, Takahishi H (2006) Stabilizing of plasmid DNA *in vivo* by PEG-modified cationic gold nanoparticles and the gene expression assisted with electrical pulses. J Cont Rel 111:382–389

Kipp JE (2004) The role of solid nanoparticle technology in the parental delivery of poorly-water-soluble-drugs. Int J Pharm 284:109–122

Koper O, Klabunde J, Marchin G, Klabunde KJ, Stoimenov P, Bohra L (2002) Nanoscale powders and formulations with biocidal activity toward spores and vegetative cells of *Bacillus* species, viruses, and toxins. Curr Microbiol 44:49–55

Kreyling WG, Semmler M, Moller W (2004) Dosimetry and toxicology of ultrafine particles. J Aerosol Med 17:140–152

Kwak K, Kim C (2005) Viscosity and thermal conductivity of copper oxide nanofluid dispersed in ethylene glycol. Korea Aust Rheol J 17:35–40

Lai MK, Chang CY, Lian YW (2006) Application of gold nanoparticles to microencapsulation of thioridazine. J Cont Rel 111:352–361

Lara HH, Ayala-Nunez NV, Turrent LI, Padilla CR (2010) Mode of antiviral action of silver nanoparticles against HIV-1. J Nanobiotech 8:1–9

Lavan DA, McGuire T, Langer R (2003) Small scale systems for *in vivo* drug delivery. Nat Biotechnol 21:1184–1191

Li B, Logan BE (2004) Bacterial adhesion to glass and metal oxide surfaces. Coll Surf B 36:81–90

Lin W, Huang YW, Zhou XD (2006) *In vitro* toxicity of silica nanoparticles in human lung cancer cells. Toxicol Appl Pharma 217:252–259

Liu Y, Li J, Qiu XF, Burda C (2007) Bactericidal activity of nitrogen-doped metal oxide nanocatalysts and the influence of bacterial extracellular polymeric substances (EPS). J Photochem Photobiol A Chem 190:94–100

Loo C, Lin A, Hirsch L (2004) Nanoshell-enabled photonics-based imaging and therapy of cancer. Technol Cancer Res Treat 3:33–40

Lu L, Sun RW, Chen R, Hui CK, Ho CM, Luk JM, Lau GK, Che CM (2008) Silver nanoparticles inhibit hepatitis B virus replication. Antivir Ther 13:253–262

Makarov VV, Love AJ, Sinitsyna OV, Makarova SS, Yaminsky IV, Taliansky ME, Kalinina NO (2014) Green nanotechnologies: synthesis of nanoparticles using plants. Acta Natur 6(1):35–44

Min JS, Kim KS, Kim SW, Jung JH, Lamsal K, Kim SB, Jung M, Lee YS (2009) Effects of colloidal silver nanoparticles on Sclerotium forming Phytopathogenic fungi. Plant Pathol J 25(4):376–380

Moghimi SM, Hunter JC, Murray JC (2001) Long circulating and target specific nanoparticles: theory and practice. Pharmacol Rev 53:283–318

Monteiro-Riviere NA, Nemanich RJ, Inman AO (2005) Multi-walled carbon nanotube interactions with human epidermal keratinocytes. Toxicol Lett 155:377–384

Niidome T, Yamagata M, Okamoto Y (2006) PEG-modified gold nanorods with a stealth character for in vivo applications. J Cont Rel 114:343–347

Nobs L, Buccheger F, Gurny R (2004) Poly (lactic acid) nanoparticles labeled with biologically active Neutravidin™ for active targeting. Eur J Pharm Biopharm 58:483–490

Oberdörster G, Maynard A, Donaldson K (2005) Principles for characterizing the potential human health effects from exposure to nanomaterials: elements of a screening strategy. Part Fibre Toxicol 2:1–8

Obordoster G (1996) Significance of particle parameters in the evaluation of exposure dose response relationships of inhaled particles. Inhal Toxicol 8:73–89

Peracchia MT, Fattal E, Desmaele D (1999) Stealth PEGylated polycyanoacrylate nanoparticles for intravenous administration and splenic targeting. J Control Release 60:121–128

Pinto DB, Shukla S, Perkas N, Gedanken A, Sarid R (2009) Inhibition of herpes simplex virus type 1 infection by silver nanoparticles capped with Mercapto ethane sulfonate. Bioconjug Chem 20:1497–1502

Rai RV, Bai JA (2011) Nanoparticles and their potential application as antimicrobials. In: Méndez V (ed) Science against microbial pathogens, communicating current research and technological advances. Badajoz Formatex, pp 197–209

RaviKumar MN, Gameti M, Mohapatra SS (2004) Cationic silica nanoparticles as gene carriers: synthesis, characterization and transfection efficiency in vitro and in vivo. J Nanosci Nanotech 4:876–881

Ravishankar V, Rai A, Jamuna B (2011) Nanoparticles and their potential application as antimicrobials. Science against microbial pathogens: communicating current research and technological advances FORMATEX 1:197–209

Ren G, Hu D, Cheng EWC, Vargas-Reus MA, Reip P, Allaker RP (2009) Characterisation of copper oxide nanoparticles for antimicrobial applications. Int J Antimicrob Agents 33:587–590

Richards R, Li W, Decker S, Davidson C, Koper O, Zaikovski V, Volodin A, Rieker T, Klabunde K (2000) Consolidation of metal oxide nanocrystals. Reactive pellets with controllable pore structure that represent a new family of porous, inorganic materials. J Am Chem Soc 122:4921–4925

Rupareli JP, Chatterjee AK, Duttagupta SP, Mukherji S (2008) Strain specificity in antimicrobial activity of silver and copper nanoparticles. Acta Biomater 4:707–771

Sadiq M, Chowdhury B, Chandrasekaran N, Mukherjee A (2009) Antimicrobial sensitivity of *Escherichia coli* to alumina nanoparticles. Nanomed Nanotech Biol Med 5:282–286

Saez A, Guzman M, Molpeceres J (2000) Freeze drying of poly-caprolactone and poly (D-L-lactic-glycolic) nanoparticles induce minor particle size changes affecting the oral pharmokinetics of loaded drugs. Eur J Pharm Bipharm 50:379–387

Sahayaraj K, Rajesh S (2011) Bionanoparticles: synthesis and antimicrobial applications, science against microbial pathogens: communicating current research and technological advances. In: Me'ndez-Vilas A (ed) FORMATEX, pp 228–244

Sarkar S, Jana AD, Samanta SK, Mostafa G (2007) Facile synthesis of silver nanoparticles with highly efficient anti-microbial property. Polyhedron 26:4419–4426

Sayes CM, Liang F, Hudson JL (2006) Functionalization density dependence of single-walled carbon nanotubes cytotoxicity in vitro. Toxicol Lett 161:135–142

Scenihr (2006) EU Scientific Comittee on emerging and newly identified health risks. Modified opinion on the appropriateness of existing methodologies to assess the potential risks associated with engineered and adventitious products of nanotechnologies. SCENIHR/002/05

Seki J, Sonoke S, Saheki A (2004) A nanometer lipid emulsion, lipid nano-sphere(LNS) as a parental drug carrier for passive drug targeting. Int J Pharm 273:75–83

Senior J, Gregoriadis G (1982) Is half-life of circulating small unilamellar liposomes determined by changes in their permeability? FEBS Lett 145:109–114

Senior J, Crawley JCW, Fisher D (1985) Influence of surface hydrophobicity of liposomes on their interaction with plasma protein and clearance from the circulation: studies with poly(ethylene glycol)-coated vesicles. Biochim Biophys Acta 1062:77–82

Sharma VK, Yngard RA, Lin Y (2009) Silver nanoparticles: green synthesis and their antimicrobial activities. Adv Colloid Interf Sci 145:83–96

Shaw IC (1999) Gold-based therapeutic agents. Chem Rev 99:2589–2600

Sheikpranbabu S, Kalishwaralal K, Venkataraman D, Eom SH, Park J, Gurunathan S (2009) Silver nanoparticles inhibit VEGF-and IL-1b-induced vascular permeability via Src dependent pathway in porcine retinal endothelial cells. J Nanobiotechnol 7:1–8

Shim J, Kang SH, Park WS (2004) Transdermal delivery of minoxidil with block copolymer nanoparticles. J Cont Rel 97:477–484

Simpson CA, Huffman BJ, Gerdon AE, Cliffel DE (2010) Unexpected toxicity of monolayer protected gold clusters eliminated by PEG-thiol place exchange reactions. Chem Res Toxicol 23:1608–1616

Soumya RS, Hela PG (2013) Nano silver based targeted drug delivery for treatment of cancer. Pharm Lett 5(4):189–197

Stern JM, Stanfield J, Lotan Y (2007) Efficacy of laser-activated gold nanoshells in ablating prostate cancer cells *in vitro*. J Endourol 21:939–943

Stoimenov PK (2002) Metal oxide nanoparticles as bactericidal agents. Langmuir 18:6679–6686

Sun RW, Rong C, Chung NPY, Ho CM, Lin CLS, Che CM (2005) Silver nanoparticles fabricated in Hepes buffer exhibit a cryoprotective activities toward HIV-1 infected cells. Chem Commun:5059–5061

Sunada K, Kikuchi Y, Hashimoto K, Fujishima A (1998) Bactericidal and detoxification effects of TiO2 thin film photocatalysts. Env Sci Tech 32:726–728

Svenson S, Tomalia DA (2005) Dendrimers in biomedical applications-reflections on the field. Adv Dru Del Rev 57:2106–2129

Taniguchi N (1974) On the basic concept of 'nano-technology' proc Intl Conf prod Eng Tokyo. Part II, Japan Society of Precision Engineering 5–10:10

The Royal Society and The Royal Academy of Engineering (2004) Nanoscience and nanotechnologies: opportunities and uncertainties. London, UK. Accessed 10 Aug 2004. http://www.nanotec.org.uk/finalReport.htm

Wei C, Lin WY, Zainal Z, Williams NE, Zhu K, Kruzic AP, Smith RL, Rajeshwar K (2004) Bactericidal activity of TiO_2 photocatalyst in aqueous media: toward a solar-assisted water disinfection system. Environ Sci Technol 28:934–938

Wist J, Sanabria J, Dierolf C, Torres W, Pulgarin C (2004) Evaluation of photocatalytic disinfection of crude water for drinking water production. J Photochem Photobiol A Chem 147:241–246

Xia T, Kovochich M, Brant J (2006) Comparison of ambient and manufactured nanoparticles to induce cellular toxicity according to an oxidative stress paradigm. Nano Lett 6:1794–1807

Xu JF, Ji W, Shen ZX, Tang SH, Ye XR, Jia DZ, Xin XQ (2009) Preparation and characterization of CuO nanocrystals. J Solid State Chem 147:516–519

Yamakoshi Y, Umezawa N, Ryu A (2003) Active oxygen species generated from photo-excited fullerene (C-60) as potential medicines: O_2−versus 1 O_2. J Am Chem Soc 125:12803–12809

Yu H, Chen M, Rice PM, Wang SX, White RL, Sun X (2005) Dumbbell like bio-functional Au-Fe $_3O_4$ nanoparticles. Nano Lett 5(2):379–382

Zhang L, Yu Y, Xiang X (2004) Camptothecin derivative loaded poly(caprolactone co lactide)-b-PEG-b-poly-(caprolactone-co-lactide) nanoparticles and their biodistribution in mice. J Cont Rel 95:135–148
Zhang LL, Jiang YH, Ding YL, Povey M, York D (2007) Investigation into the antibacterial behaviour of suspensions of ZnO nanoparticles (znonanofluids). J Nanopart Res 9:479–489
Zharov VP (2006) Photothermalnanotherapeutics and nanodiagnostics for selective killing of bacteria targeted with gold nanoparticles. Biophys J 90:619–627

Chapter 8
Enzymology and Nanoparticles

Sumera Javad and Khajista Jabeen

8.1 Introduction

Enzymology is the study of enzymes and the reactions catalyzed by the enzymes. The field of enzymology started in early nineteenth century and in 1833, Payen and Persoz first recognized the enzyme which was involved in sugar break down. Pepsin enzyme was first isolated by Schwan in 1834 from an animal source by acidification of stomach wall. The term enzyme was first coined by Kuhne in 1879 and Sumner (1926) firstly obtained urease from jack bean. History of this word has Greek origin as well as German. In Greek "en" means "within" while "zume" means leaven, so word is enzymos which in late nineteenth century came to be enzyme, a German word.

Enzymes are biological catalyst made up of proteins that boost the speed of biochemical reactions without being altered themselves. Basically the enzymes reduce the activation energy of a reaction upto 10^8 folds. All enzymes are large in size ranging from 14,000 to 400,000 Da which are water soluble and remain tightly bound to cell membranes. The little quantity of enzymes are required in a reaction, depending upon their characteristics like pH, substrate quantity and incubation period. Each enzyme has an active site where substrate binds. When enzyme acts on substrate, the substrate is bound to the active site of the enzyme and an ES (enzyme substrate) complex is formed which is converted into P (product). Enzyme activity is the amount of enzymes utilized in a reaction to convert the substrate into product and it is measured in min mL^{-1}.

Enzymes reduce the activation energy of the reaction which is required to initiate a reaction. Thus making it rapid and consuming lesser energy as a whole. Due to which a number of otherwise impossible reactions are being carried out in the biological systems with the help of enzymes (Fig. 8.1).

S. Javad · K. Jabeen (✉)
Department of Botany, Lahore College for Women University, Lahore, Pakistan

S. Javad, A. Butt (eds.), *Nanobotany*, https://doi.org/10.1007/978-3-319-77119-9_8

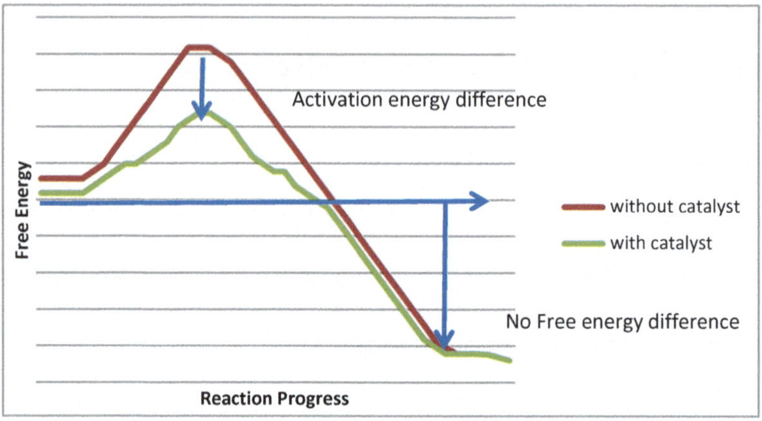

Fig. 8.1 Reaction energies of catalyst assited and non-catalyst assisted reactions

The activity of an enzyme also depends on some compounds known as inhibitors and activators. The inhibitors inhibit the reaction catalyzed by enzymes which might be reversible and irreversible. The reversible inhibition further classified as non-competitive inhibition, competitive inhibition and uncompetitive inhibition. In competitive inhibition a substance analogues to substrate is competing with the substrate to bind with the active site, whereas non-competitive and un-competitive inhibition is un-successful. A cofactor (non-protein part of enzyme) is also required sometimes for enzyme activation. The cofactors are both inorganic and organic in nature; mostly organic co-factors are also called as co-enzymes which are generally vitamins or vitamin derived compounds. Up till now 3000 enzymes are discovered and classified in six major categories including Transferases, Oxidoreductases, Hydrolases, Ligases, Lyases and Isomerases.

The enzymes molecules aggregates increase the surface area of enzymes and these aggregate when reached to 100 nm are known as enzyme nanoparticles (ENPs). Enzyme nanoparticles are specialized enzyme structures having specific protein configuration in nanometer diameter ranges from 10–100 nm.

8.2 Nanotechnology

A branch of science that deals with matter at the scale of 1 billionth of a meter (i.e., 10^{-9} m = 1 nm) is known as "Nanotechnology". Nanoparticles can be defined as objects ranging in size from 1–100 nm that act as a complete entity in terms of its carriage and properties. Metallic nanoparticles possess diverse chemical and physical characteristics from bulk metals (e.g., higher specific surface areas, lower melting points, specific optical properties, specific magnetizations and mechanical strengths) that might be attractive for several industrial applications.

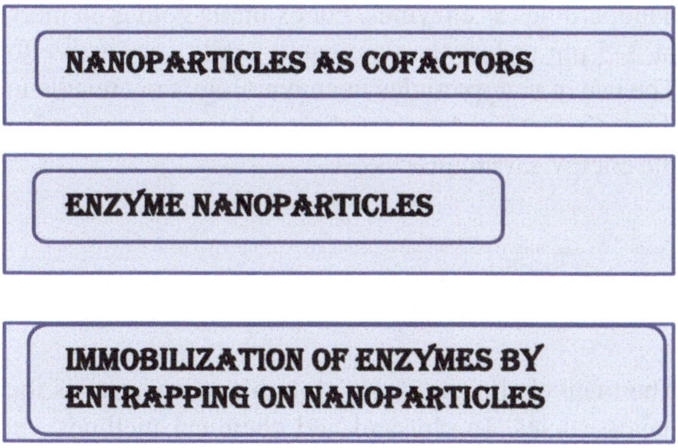

Fig. 8.2 Enzymology and nanotechnology

Nanotechnology is not so recent, people before fourth century used optical characteristic of nanoparticles for preservation of paintings and sculptures. The practice of using nanoparticles in medical field is also evident from history for the cure of epilepsy, tumors, dysentery and heart diseases. However, Daniel and Astruc first time compiled the nanoparticles history from middle ages to ancient times. In 1618 Francisci Antonii published a book on gold colloidal particles. Kunckel (1676) first time industrially manufactured the stained glass with the help of colloidal particles. Subsequently, an era of inventive progress in the field of nanoparticles started in 1970s. This boost was possible due to combine efforts of researchers from various fields of science including biosciences, chemistry, physics and material sciences.

Nanoparticles applications in various arenas necessitate a simple and inexpensive mode for the synthesis of high value nanoparticles. In recent year significant attempts have been made to synthesize nanoparticles from plant extract by microwave radiations. All this growing knowledge about the synthesis and characteristics of nanoparticles is also helpful in expanding the practical applications of enzymes in industry. Combination of nanotechnology and enzymology is being used for Enzyme nanoparticles, immobilization of enzymes or enzyme nanoparticles and use of nanoparticles as cofactors of enzymes (Fig. 8.2).

8.3 Nanoparticles as Cofactors

There are number of various commercial applications of biocatalysts/enzymes in different fields of research and industries. It includes food industry, chemical industry and agricultural industry. To increase the efficiency of these reactions different additives are used. These are selected for their specificity and stability. The concept of real supported catalyst is also in application. Nanoparticles of controlled size, shape and composition can play a great role in the field of catalysis processes which are involved in different industrial productions. Particle size, shape and supported materials are the main factors which help in the stability and selectivity of nanoparticles as enzymes. For example gold is an inert material but its nanoparticles of 3–5 μm or less than are catalytically very active for several chemical reactions. The use of nanoparticles in enzymology may results in a large contact area between the active site and surrounding substrate material. It will be helpful in developing the energy saving processes.

8.3.1 Nano-surfaces as Cofactors: Challenges

The main challenge involved in this perspective is the control of size and shape of nanoparticles. In physical and chemical methods, metal oxide support on which nanoparticles are grown plays a vital role. In green synthesis, there are also different parameters which can be optimized to control the size and shape of nanoparticles. This can be clearly understood with the example of development of one dimensional gold nanoparticles chains on alumina and two dimensional islands of gold nanoparticles on MgO films, provided all other conditions same (Nilius et al. 2008; Sterrer et al. 2007). The basic principle lies in the transfer of charges in between the layers. Sometimes dopants are introduced for this purpose e.g., 3D layers and 1D layers of gold nanoparticles are formed on CaO and dopped CaO layers respectively (Shao et al. 2011).

Schauermann et al. (2012), also confirmed that decreasing the particle size, increase the efficiency of the catalyst. But it is not the ultimate reason. Nature of particles used and their nanoclusters also play their role. They further added that lattice surface of nanoparticles is lesser as compared to normal metals. It results in the decrease in the binding energies. Contracted clusters cause weaker adsorption bonds and stronger binding energy.

Hvolbaek et al. (2007) also reported the effect of particle size of gold nanoparticles. They said that lower size nanoparticles decrease the temperature of reaction significantly. Infact nanoparticles present an increased surface area of contact between reactants and catalysts.

8.3.1.1 Molbdinium as Cofactor

The trace element molybdenum is essential for nearly all organisms and forms the catalytic center of a large variety of enzymes such as nitrogenase, nitrate reductases, sulphite oxidase and xanthine oxidoreductases. Nature has developed two scaffolds holding molybdenum in place, the iron–molybdenum cofactor and pterin-based molybdenum cofactors. Despite the different structures and functions of molybdenum-dependent enzymes, there are important similarities, which we highlight here. The biosynthetic pathways leading to both types of cofactor have common mechanistic aspects relating to scaffold formation, metal activation and cofactor insertion into apoenzymes, and have served as an evolutionary 'toolbox' to mediate additional cellular functions in eukaryotic metabolism (Schwarz et al. 2009).

There are a number of various commercially exploited biocatalysts or enzymes which are being used in different industries and they need different metal cofactors.

(i) Cellulases

Cellulases enzyme is used in various industries like agriculture, brewing and wine making, food, textile, animal feed, biomass refining, laundry and pulp and paper industry (Kuhad et al. 2011). Mg^{2+} Ions are used to enhance the activity of Cellulase enzyme (Naseeb et al. 2015).

(ii) Lipases

Lipases act on carboxylic ester bonds and they are part of Hydrolases group of enzymes. Lipases enzymes have a very long list of industrial application. These enzymes are used in detergent industry, in cosmetics, pharmaceuticals, food, paper industry, textile and leather (Houde et al. 2004).

Lipases enzymes are group of enzymes in which activity of some members is enhanced by iron and cobalt (Fetzner 2002). There are also research reports showing that activity of lipases is enhanced by the action of alkaline earth metals as cofactors (Ota and Yamada 1966). This suggests the use of iron and cobalt nanoparticles as cofactors to increase the activity of Lipases.

(iii) Alpha-amylases

This is a large group of enzymes used to convert larger units of polysaccharides into smaller units. They cause the hydrolysis of internal α-1,4-glycosidic linkages of starch in low molecular weight products such as maltose, glucose and maltotriose units. These have a very vast application in the industry such as food, textile, fuel alcohol production, pharmaceutical, fermentation, paper and detergent industry. These form the 25% of the total enzyme industry. The human α-amylase, a domain by disulphide bond. Ca^{2+} is required for the stabilization of this structure of amylases as well as for its activation (Haq et al. 2010; Souza 2010).

(iv) Proteases

Proteases are used on large scale in textile industry. They are also used in the sea food industry. The history of their use in detergent industry goes back to 1913, which became modernized with time. They are also used in dishwashers. Their main purpose is to remove proticnacious spots from fabric or utensils (Haard and Simpson 1994; Maurer 2004). It has been reported that manganese and cobalt are required for the activation of peptidases (Wang et al. 2003).

(v) Glucanases

These enzymes are used in brewing and malt industry. In these industries high molecular weight glucans of cell walls cause a great problem as they increase the viscosity of the product. This causes the problem with the whole system such as pumping, flowing etc. Application of external glucanases hydrolyze the high molecular weight glucanes and thus aiding in the industrial process as well as making the quality better (Celestino et al. 2006). In poultry, interest in the use of glucanases enzyme for non-viscous cereal grains has increased (Choct 2006).

8.4 Immobilization of Enzymes on Nanosurfaces

Stability of enzyme can be easily increased by immobilizing them on some surfaces and use of nanoparticles for this purpose increases the advantages. Different forms of nanoparticles can be used for this purpose like nanoparticles, nanofibers and mesoporous silica etc., small particles can provide a large surface area for the attachment of the enzyme molecules. For example the use of magnetic nanoparticles of iron oxide as a immobilizing surface for the molecules of lipsase by Dyal et al. (2003). Enzyme molecules showed stability upto 1 month without any change in the activity. Nanofibers further improved the process by providing more surface for the enzyme attachment. It was also suggested that fibrous nanofibers can improve the process by decreasing the diffusional path for substrate molecules (Jia et al. 2002). Mesoporous silica is the material with high pore volume, high surface area and well-ordered structured pores (Beck et al. 1992).

8.5 Enzyme Nanoparticles (ENPs)

Enzyme Nanoparticles are an assemblage or aggregate of enzyme molecules in a static system of protein configuration between 10–100 nanometer (nm) planes.

Enzymes possess great potential as healing mediators due to their distinctive highly active specificity. Current advances in the synthesis and characterization of nanoparticles for their unique characteristics enhance the possible use of nanoparticles as carriers of enzyme medicines.

ENPs (enzyme nanoparticles) have distinctive optical, thermal, electrical, catalytic, mechanical and chemical properties. These ENPs have capacity to tailor the nanoparticles that serve to enhance the performance of enzyme based sensors. Theses ENPs directly attached on nanoparticles caused the denaturation of nanoparticles. To avoid denaturation of nanoparticles after attachment with enzymes, enzymes arranged themselves in clusters in an organized cross linked sequence. This strategy of ENPs made easy the development of biosensors (Sharma et al. 2011). A biosensor is a device which converts an observed biological response into a measurable electric signal. The magnitude of this signal is directly proportional to the intensity of biological response e.g., the concentration of the chemical in the sample etc.

After this advancement, many researchers prepared ENPs biosensors for Silver nitrate, uricase, cholesterol and glucose oxidase (Kundu et al. 2012; Chawla et al. 2013; Chauhan et al. 2014).

There is a continuous urge to use enzyme with higher performance and more stability in industrial environment which is usually harsh due to high temperatures and use of organic solvents etc. One of the approaches still used, is to encapsulate the enzymes. But this technique decreases the enzyme substrate reactivity. Then the solution came in the form of encapsulation of single enzyme nanoparticles (SENs) which are different from ENPs (Yan et al. 2006). SENs are single enzyme

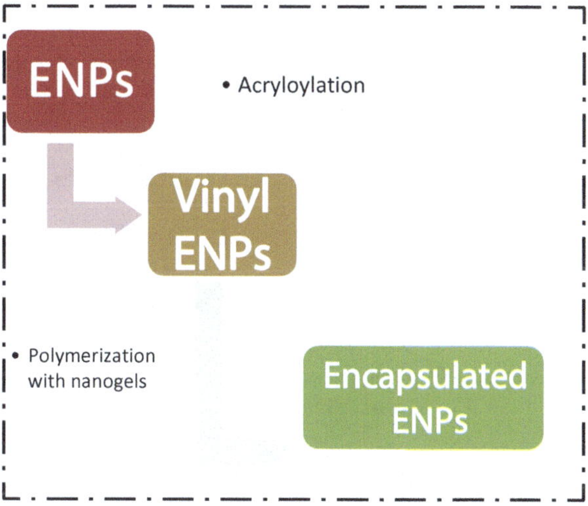

Fig. 8.3 Synthesis of SENs

nanoparticles bounded by an organic or inorganic nanoscale macromolecules network. By using SENs technique many SENs of trypsin, cellulase and chymotrypsin were prepared by Kim and Grate (2003) and Blanchette et al. (2012). This technique makes enzymes stable and more reactive for the industrial scale.

Its not even a very complicated process. For example encapsulation of single enzyme nanoparticles in nanogel, involves two steps,

(i) Acryloylation

Its purpose is to produce vinyl groups on the surface of protiens.

(ii) In situ polymerization

In this step aqueous in situ polymerization is done with the nanogels. This step may be repeated several times with different monomers or crosslinkers to get the SENs of desired shape and size. These coating materials are permeable to substrate as well as product (Fig. 8.3).

8.5.1 Preparation of ENPs

ENPs can be prepared by addition of ethanol (drop wise) at 4 °C, this causes hydration of cell wall of enzyme molecules. Various factors are included like hydrophobic, van der Waals and electrostatic forces subsequently to form ENPs. For the formation of ENPs Cross linking is the simplest option. In this process different biocatalyst or reagents react with each other to form ligands through covalent bonding (Datta et al. 2013). Glutaraldehyde is commonly used as a functional cross-linking agent.

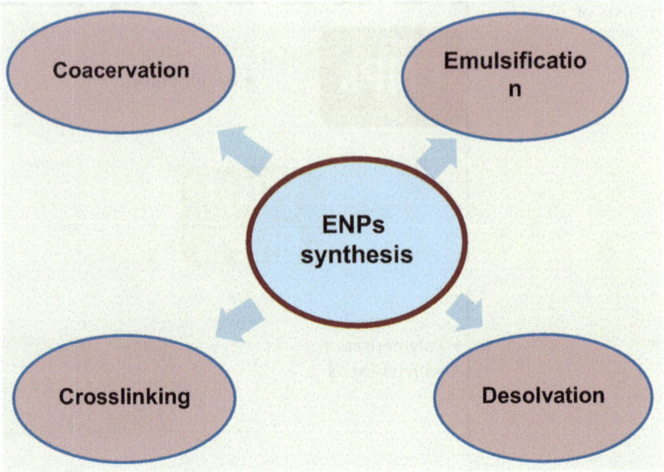

Fig. 8.4 Methods to produce ENPs

As enzymes are protein in nature so in the formation of ENPs, the nature of proteins (whether they are inert or soluble proteins like albumin and gelatin) also played important role and can be useful either to cell or enzymes (Fig. 8.4).

These proteins can be used in the formation of ENPs by following methods:

(1) Emulsification
(2) Desolvation
(3) Simple coacervation
(4) Cross-linking in water and oil emulsion

(1) **Emulsification**

In emulsification aqueous albumin solution (A standard protein) is converted into emulsion at room temperature by adding cotton seeds oil (Vegetable oil) using a high speed mechanical homogenizer. The obtained emulsion is added drop by drop in preheated oil at 120 °C which results in prompt evaporation of water content and albumin molecules and NPs generated in this way (Scheffel et al. 1972).

(2) **Desolvation**

ENPs have also been formed by desolvation method with ethanol and successive cross-linking with glutaraldehyde. In this process 1–2 mg mL^{-1} of enzyme is dissolved in reaction buffer. Ethanol is added drop wise (0.1–0.5 mL min^{-1}) in dissolved enzyme in 1:2 ratio at 4 °C in a stirrer at 500–800 rpm. Ethanol (desolvating agent) reduces the space between enzyme molecules by removing water molecules and boost up the enzyme molecules to form small clusters of 1–100 nm size. 1.8 mL of glutaraldehyde (2.5–8.0%) is also added into 9 mL suspension buffer of ENPs for 24 h at 4 °C in a stirrer at 500–800 rpm in an ice bath. This process caused cross-linking of ENPs (Liu et al. 2005).

According to Pundir (2015) Schiff base formation involved in the formation of ENPs. According to this method from gulataraldehyde, two aldehyde groups become attached with the free NH_2 group of lysine, thus forming residues on the external surface of ENPs. The addition of Cysteamine in suspension of ENPs leads towards the release of NH_2 group and Cysteine gives _SH group to surface of ENPs.

(3) **Simple coacervation**

In simple coacervation method in 150 mL BSA (5 mg L^{-1} in 10 mM Tris/HCl comprising 0.02% sodium azide at 7.5 pH) ethyl alcohol was added to make a turbid solution. This solution was cross-linked with 25% glutaraldehyde (150 μL) at 24 °C, after this ethanolamine was added to block aldehyde. Tween-20 [0.01% (v/v)] is then added as stabilizer. Large size particles are removed through centrifugation at 4 °C for 30 min. This supernatant was dialyzed and ultra-filtered through copolymer polyvinyl chloride membrane and 0.2 μm acetate membrane respectively.

(4) **Cross-linking**

The emulsification of BSA was done by its homogenization at high pressure in pre heated oil at 100 °C. This process denatured the protein content and allowed water to evaporate and resulted in the formation of ENPs. These ENPs can be purified by centrifugation at 4 °C for 10 minutes at 14000 rpm and the pellets were dispersed through 5 minutes ultrasonication.

8.5.2 How Enzyme Nanoparticles Bind with Substrate?

Enzymes attached themselves to a substrate either physically or chemically to maintain its activity. A term immobilization is commonly used for this purpose. In this process ENPs covalently bound themselves to a substrate through covalent binding, cross-linking, entrapment and adsorption. In covalent binding, covalent bonds formed between functional groups of ENPs and supporting medium. According to D'Souza in (1999) covalent link of enzymes with supporting medium happens due to the side chain of various amino acids e.g. aspartic acid, arginine, histidine and extent of their reactions with various functional groups including hydroxyl, indolyl and imidazole etc. Cross linking involved the covalent bonding of various biocatalyst and chemical reagents to form ligands. Entrapment is a process in which the movement of ENPs is restricted to a porous material like agrose gel, agar and gelatin. Adsorption of ENPs is a mechanism by which ENPs can be immobilized onto insoluble matrix/medium. Adsorption mechanism is established on hydrogen, ionic bonding and vander Waal interaction.

These bondings help ENPs to be reused many times and being functional for a long time period. The nitrocellulose membrane (NC), Au electrode and Pt eclectrodes are commonly used to immobilize ENPs (Sharma et al. 2011; Kundu et al. 2012; Chawla et al. 2013).

For example Chauhan et al. (2014) suggested a protocol to immobilize Thiolated uricase nanoparticles aggregates onto Au electrode covalently. According to his findings firstly Au electrode was cleaned and scanned with 0.2 M H_2SO_4 over a range of 1.5–0.5 V, the procedure was continued unless the characteristic polycrystalline Au is established on a voltammogram. The polycrystalline Au electrode was placed in the uricase (NPs) suspension at 4 °C for 12 h and stirred slightly. The Au electrode bounded with self-assembled uricase nanoparticles was rinsed with 50 mM of phosphate buffer having 7.4 pH. The obtained Au electrode with nanoparticles was stored in same buffer and placed at 4 °C for further usage.

Kundu et al. (2012) also had given a protocol to immobilize glucose oxidase (GOD) nanoparticles on nitrocellulose membrane with Pt electrode. In this procedure the Nitrocellulose membrane with chitosan or Pt electrode with cysteamine HCl introduce _NH_2 groups on their surfaces. Enzyme nanoparticles were immobilized covalently on amino functionalized nitrocellulose membranes via gluteraldehyde coupling (Fig. 8.5).

In this process reaction occurs on Nitrocellulose membrane between amino group of membrane and aldehyde group of gluteraldehyde on one end. The free _ CHO group of glutaraldehyde on its other end reacts with _NH_2 groups on the surface of ENPs aggregates. This formed the Schiff base between glutaraldehyde-

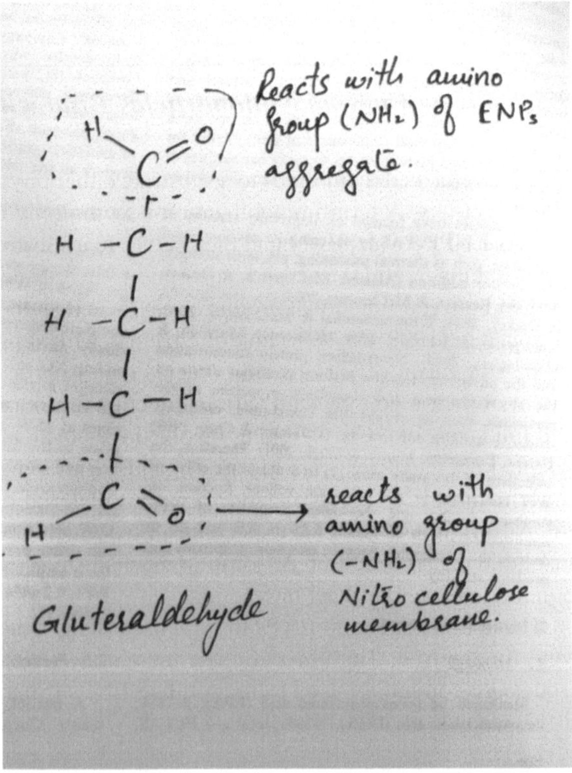

Fig. 8.5 Mechanism of action of gluteraldehyde

activated Nitrocellulose membrane/Pt electrode and ENPs aggregates. The immobilization of these ENPs was confirmed through SEM (scanning electron microscope).

8.5.3 Kinetics of ENPs

ENPs showed enhanced stability due to higher surface area but the kinetic properties of ENPs not yet reported in literature. Although due to immobilization of ENPs through covalent bonding on metals like (Au & Pt electrode) and nitrocellulose membrane made bit easier to study the kinetic properties of ENPs.

As we know that temperature, pH and time of contact between substrate and enzyme are necessary factors which affect the kinetics of enzyme and in return enzyme nanoparticles complex. The immobilization positively affects the activity of ENPs against high temperature regimes as compared to free enzyme, for example 5 °C is an optimum temperature for glucose oxidase (Kundu et al. 2012). The immobilization and cross linking provide thermostability to ENPs which made them stronger against conformational changes due to high temperature.

Slight increase in optimum pH also increased the activity of ENPs when compared to free enzymes. According to many scientists an increase in pH increased the stability of ENPs (Liu et al. 2005; Sharma et al. 2011). It has been proved from experimentation that slight change of these controlling factors like pH, temperature can affect the characteristics of ENPs on larger scale as compared to simple enzymes. Micronization of substrate also affects the enzymatic activity of ENPs (Wu et al. 2000).

Immobilization of ENPs is pretty much helpful to reuse the enzyme particles again and again. It also makes them more stable. Enzymatic reactions can be easily controlled as supporting plates can be moved in and out. Different enzymatic models can be prepared with the help of supporting plates like designing multi-enzyme system. We can get a more pure product without mixing of any other molecule industrially as well as the complete usage of the reactants. These characteristics make ENPs a guarantee to the economic success of the industry.

8.5.4 Applications of ENPs

Currently ENPs are commercially used in the fabrication of highly sensitive, low cost, and environmental friendly products like:

Biosensors production
Biomedical devices
Mini biochips
Biofuel cells
Enzyme reactors

8.5.5 Future Development

The challenging study in the field of ENPs production is sample preparation. Concentrated acid solutions were traditionally used for the extraction of enzyme metallic particles (ENPs) for complete dissolution of biological and environmental samples (like water, sediments, clay, inorganic salts, complex matrices, metal oxides and organic debris).

For better understanding of transport, exposure and toxicity of ENPs depends solitary on the specific and exact characterization of ENPs. So the future research should specifically be focused on Nano metrology.

References

Beck JS, Vertuli JC, Roth WJ, Leonowicz ME, Kresge CT, Schmitt KD, Chu CTW, Olson DH, Sheppard EW, Mccullen SB, Higgins JB, Schelker JL (1992) A new family of mesoporous molecular sieves prepared with liquid crystal templates. J Am Chem Soc 114:10834–10843

Blanchette C, Lacayo CI, Fischer NO, Hawang M, Thelen MP (2012) Enhanced cellulose degradation using cellulose nanospheres. PLoS One 7:e42116

Celestino KRS, Cunha RB, Felix CR (2006) Characterization of a β-glucanase produced by rhizopus microsporus var. Microsporus, and its potential for application in the brewing industry. BMC Biochem 7(1):1

Chauhan N, Kumar A, Pundir CS (2014) Construction of an uricase nanoparticles modified Au electrode for amperometric determination of uric acid. App Biochem Biotechnol 174:1683–1694

Chawla S, Rawal R, Sonia R, Pundir CS (2013) Preparation of cholesterol oxidase nanoparticles and their application in amperometric determination of cholesterol. J Nanopart Res 15:1934–1943

Choct M (2006) Enzymes for the feed industry: past, present and future. Worlds Poult Sci J 62(01):5–16

D'Souza SF (1999) Immobilized enzymes in bioprocess. Curr Sci 77:69–79

Datta S, Christena LR, Rajaram YRS (2013) Enzyme immobilization: an overview on techniques and support materials. Biotech 3:1–9

Dyal A, Loos K, Noto M, Chang SW, Spagnoli C, Shafi K, Ulman A, Cowman M, Gross RA (2003) Activity of *Candida regusa* lipase immobilized on gamma Fe_2O_3 magnetic nanoparticles. J Am Chem Soc 125:1684–1685

Fetzner S (2002) Oxygenases without requirement for cofactors or metal ions. Appl Microbiol Biotech 60(3):243–257

Haard N, Simpson B (1994) Proteases from aquatic organisms and their uses in the seafood industry. Fisheries processing. Springer, pp 132–154

Haq I, Ali S, Javed M, Hameed U, Saleem A, Adnan F, Qadeer M (2010) Production of alpha amylase from a randomly induced mutant strain of *Bacillus amyloliquefaciens* and its application as a desizer in textile industry. Pak J Bot 42(1):473–484

Houde A, Kademi A, Leblanc D (2004) Lipases and their industrial applications. Appl Biochem Biotech 118(1–3):155–170

Hvolbæk B, Janssens TVW, Bjerne SC, Hanne F, Claus HC, Jens KN (2007) Catalytic activity of Au nanoparticles. NanoToday 2(4):14–19

Jia HF, Zhu GY, Vugrinovich B, Kataphinan W, Reneker DH, Wang P (2002) Enzyme carrying polymeric nanofibers prepared via electrospinning for use as unique biocatalysts. Bitech Prog 18:1027–1032

Kim J, Grate JW (2003) Single enzyme nanoparticles armored by a nanometer scale organic/inorganic network. Nano Lett 3:1219–1222

Kuhad RC, Gupta R, Singh A (2011) Microbial cellulases and their industrial applications. Enzyme Res Article ID 280696, 10 pages

Kundu N, Yadav S, Pundir CS (2012) Preparation and characterization of glucose oxidase nanoparticles and their application in DO metric determination of serum glucose. J Nanosci Nanotechnol 13:1710–1716

Liu G, Lin Y, Ostatna V, Wang J (2005) Enzyme nanoparticles based electronic biosensor. Chem Commun (27):3481–3483

Maurer KH (2004) Detergent proteases. Curr Opin Biotech 15(4):330–334

Naseeb S, Sohail M, Ahmad A, Khan SA (2015) Production of xylanases and cellulases by Aspergillus fumigatus ms16 using crude lignocellulosic substrates. Pak J Bot 47:779–784

Nilius N, Ganduglia-Pirovano MV, Brazdova V, Kulawik M, Sauer J, Freund HJ (2008) Counting electrons transferred through a thin alumina film into Au chains. Phys Rev Lett 100

Ota Y, Yamada K (1966) Lipase from candida paralipolytica: Part ii. Alkaline earth metal ions as the cofactor in the shaken system containing no emulsifier. Agr Biol Chem 30(10):1030–1038

Pundir C (2015) Enzyme nanoparticles. Preparation, characterisation, properties and applications. eBook ISBN: 9780323389297, Elsevier

Schauermann S, Nilius N, Shaikhutdinov S, and Freund HJ (2013) Nanoparticles for Heterogeneous Catalysis: New Mechanistic Insights. Acc Chem Res 46 (8): 1673–1681

Scheffel U, Rhodes BA, Natrajan TK, Wagner HN (1972) Albumin microspheres for study of reticuloendothelial system. J Nucl Med 13:498–503

Schwarz G, Mendel RR, Ribbe MW (2009) Molybdenum cofactors, enzymes and pathways. Nature 460(7257):839–847

Shao X, Prada S, Giordano L, Pacchioni G, Nilius N, Freund HJ (2011) Tailoring the shape of metal ad-particles by doping the oxide support. Angew Chem Int Ed 50:11525–11527

Sharma S, Shrivastav A, Gupta N, Srivastava S (2011) Amperometric biosensor: increased sensitivity using enzyme nanoparticles. In: 2010 International conference on nanotechnology and biosensors. IPCBEE 2:2426

Souza PMD (2010) Application of microbial α-amylase in industry-a review. Braz J Microbiol 41(4):850–861

Sterrer M, Risse T, Heyde M, Rust HP, Freund HJ (2007) Crossover from three dimensional to two-dimensional geometries of Au nanostructures on thin MgO (001) films: a confirmation of theoretical predictions. Phys Rev Lett 98:206103–206101

Wang J, Sheppard GS, Lou P, Kawai M, Park C, Egan DA, Schneider A, Bouska J, Lesniewski R, Henkin J (2003) Physiologically relevant metal cofactor for methionine aminopeptidase-2 is manganese. Biochemistry 42(17):5035–5042

Wu C, Jim TF, Gan Z, Zhao Y, Wang S (2000) A heterogeneous catalytic kinetics for enzymatic biodegradation of poly(ϵ-caprolactone) nanoparticles in aqueous solution. Polymer 41(10):3593–3597

Yan M, Jun G, Pingkai O (2006) Encapsulation of single enzyme in nanogel with enhanced biocatalytic activity and stability. J Am Chem Soc 128:11008–11009

Chapter 9
Synthesis of Nanoparticles by Microbes

Arusa Aftab

9.1 Introduction

Microorganisms are making and running the world. Animals and plants are dependent upon micro-organisms for their life and regular metabolism. Our individual cell on average is hosting 10 microbes which are helping in vitamin synthesis, digestion and many more. They are also involved in environmental protection. They are playing their role in agriculture by increasing the soil fertility (biocomposting, bioleaching) and even as biopesticides, and now going to new horizons is their role in Nanotechnology (Kumar and Gopal 2015; Stark 2010).

Nanotechnology is the branch of technology dealing with the particles sizing 1–100 nm for at least one of their dimensions. But with the passage of time this is going to be an interdisciplinary field with a great impact in all practical fields. Its having positive role in impacting the fields of medicine, electronics, energy and space sciences. In the last decade, the interest in these fields has increased on a faster rate (Azam et al. 2009).

There are a number of methods available for the synthesis of nanoparticles. Methods already available for synthesis of nanoparticles are chemical, physical, biological or hybrid methods. Physical and chemical methods are more famous and applicable, but they pose threat by adding chemicals and heat to the environment. Synthetically nanoparticles are produced by the chemicals which are otherwise strong reducing agents which are toxic to environment and living beings. So there is a real need of use of some alternative methods which are safer for living beings and greener to the environment. The nanoparticles such prepared can't be applied in biomedical and clinical applications. So it's much significant issue to develop a method for the synthesis of nanoparticles which is independent of toxic materials and is ecofriendly. In this area of research the use of microorganisms seems

A. Aftab (✉)
Department of Botany, Lahore College for Women University, Lahore, Pakistan

S. Javad, A. Butt (eds.), *Nanobotany*, https://doi.org/10.1007/978-3-319-77119-9_9

promising. Current research is emphasizing on biological pathways like by using microorganisms, plant extracts, and fungi to produce nanoparticles (Abou El-Noor et al. 2010; Daniel and Astruc 2004).

Researches are there to show that microorganisms can synthesize nanoparticles very efficiently both inside the cell as well as outside of the cell environment. These include bacteria, fungi yeast and Actinomycetes. Research projects are still going on the understanding of the mechanism of microbial synthesis of nanoparticles with special emphasis on their practical applications. It is also gaining interest because it's cheaper and not time consuming. Furthermore by the development of modern fermenter technology it's possible to control the microbial growth in all way.

Literature shows the synthesis of a number of classes of nanoparticles use microorganisms as synthetic machinery. These examples include Platinum, gold, silver, zinc oxide, cadmium sulphide, magnetite, silica zirconia, titanium and cadmium selenide quantum dots. There are a number of natural nanoparticles-materials present. They are present in earth crust from the very start of the evolution. Environmental factors have reducing agents as well as reducing energies, so supporting the production of nanoparticles continuously. These natural nanoparticles materials include silicates, hydroxides and oxides of iron, aluminium and silicon etc in the form of clay and rocks.

The origin of all multicellular life is from nanosized organic molecules. Then the knowledge was translated to define the nanomaterials and to define the branch of technology as nanotechnology. Nanotechnology is basically the science of production, designing, characterization and practical applications of the nanoscale structures, devices or systems (Byrappa et al. 2008).

Microbes produce nano-materials as part of their digestion and are further used for different applications. Some scientists have also named them as micro-factories or bio-factories of nanoparticles. On one side of the scenario, nanoparticles are attracting the researchers while on the other side, biotechnology and microbiology have joined the hands with nanoparticles. Microbiology has a very much practical application to apply the microbes for the synthesis of nanoparticles.

The first evidence of the synthesis comes from a bacterial strain named as *Pseudomonas stutzeri* AG259, originally isolated from silver mine. There are different mechanisms proposed for the synthesis of nanoparticles by microbes. The most common is the defense mechanism of microbes itself. For example silver ions are toxic for bacterial cells when present in their environment. Bacterial cells have developed resistance system against these which is basically the bio machinery for synthesis of nanoparticles. Bacterial cells transform silver ions into silver atoms and tend to secrete them outside and accumulate them in their periplasmic space. Some bacterial strains like *E.coli* can synthesize silver nanoparticles as a part of their regular metabolism when they are in contact with the silver ions from their surroundings (Luechinger et al. 2010).

There are number of culture conditions which can be controlled for the optimized synthesis and recovery of the nanoparticles by microorganisms. One of the very important factors is temperature which plays its role by controlling the size of the particles. Temperature of the medium and size of particles are inversely related that means by increasing the temperature of the microbial culture we get small sized particles. For example nanoparticles of size 50 nm are produced at room temperature from microbial culture maintained at room temperature. But as the temperature is increased up to 60 C the size of nanoparticles is decreased to 15 nm. Other than temperature pH is also an important factor which plays its role. Nanoparticles are synthesized more efficiently under alkaline condition of medium and compared to acidic or low pH medium. But very high pH like more than 10, cell death starts, so there is reduction of nanoparticles synthesis rate. The concentration of the raw material in the environment is also an important factor. Rate of nanoparticles synthesis is influenced by change in silver by change of the silver nitrate conc in the surroundings (Liu et al. 2011; Luechinger et al. 2010).

9.2 Types of Nanoparticles Synthesized by Microbes

9.2.1 Platinum Nanoparticles

Platinum nanoparticles (PtNPs) have received the much interest of scientists in their use as catalysts, catalyst sensors and adsorbents. Nanoparticles crystals of Fe-Pt are found to have anti-cancerous activity [a-7, a-10]. Chemical methods of PtNPs synthesis require too much heating or use of strong chemicals like HCl in case of electrochemical deposition. While microbes can do the same job at room temperatures and almost at neutral pH (Tan et al. 2003; Dean 1979). *Desulfovibrio desulfuricans* (a sulfate reducing bacterium) and *Shenavella oniedensis* (a metal reducing bacterium) are capable of reducing platinum ions into PtNPs. *S. algae* is also capable of depositing the PtNPs at its surface at room temperature, pH 7 and within almost 60 min (Konishi et al. 2007). Riddin et al. (2010) reported that ratio of metal to the microbial proteins determine the morphology of the PtNPs which can be controlled. They also reported that PtNPs are deposited in the periplasmic space.

9.2.2 Gold Nanoparticles

Gold nanoparticles (AuNPs) have a place in history and practical applications of science, for examples from the history of old Roman cultures it is evident that the use of gold nanoparticles was very common there to stain glasses for refreshment feelings. They also used AuNPs in medicines for curing diseases. Michael Faraday is believed to be the pioneer of the advanced period of the use of AuNPs blend and he started more than 150 years back (Zhang et al. 2011). The extracellular synthesis of gold nanoparticles has been reported by a number of microbiologists (Hayat 1989). *Rhodococcus sp.* is observed to prepare gold nanoparticles of uniform shape. But it requires extreme environmental parameters like elevated temperatures in acidic or alkaline situations (Ahmad et al. 2003a, b; Mukherjee et al. 2001b) (Fig. 9.1).

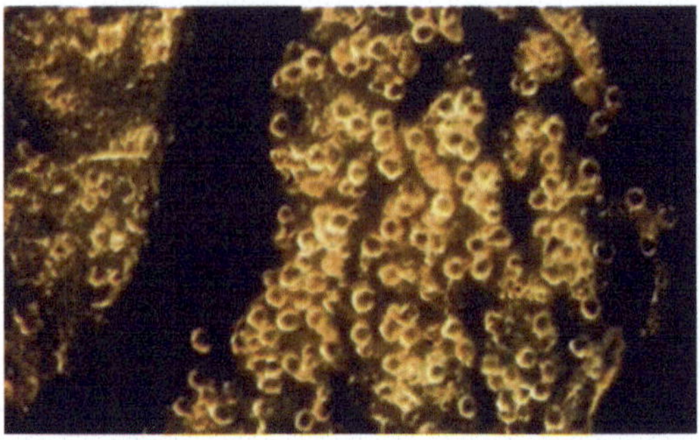

Fig. 9.1 Gold nanoparticles (Source: http://www.inquisitr.com/1310432/liposuction-goes-nano-gold-nanoparticles-could-melt-away-fat/)

Lengke et al. (2006a) reported that the filamentous cyanobacteria produce different shapes of gold nanostructures like circular, octahedral and cubic etc.

9.2.3 Silver Nanoparticles

There is a very strange relationship between microorganisms and the nanoparticles. As for example silver nanoparticles have proved to be very effective antimicrobial agents against a number of Gram –ve as well as Gram +ve bacterial strains (Lengke et al. 2006b). At times they are the most exploited nanoparticles in the fields of textile, pharmaceutical industries and agriculture as an antiseptic agent. Meanwhile they are the biofactories for the synthesis of nanoparticles. *Pseudomonas stutzeri, Pseudomonas aeruginosa, Bacillus cereus, B. subtilis, E. coli, Bacillus licheniformis, Bacillus amyloliquefaciens, Pseudomonas stutzeri AG259, Phaeocystis antarctica, Pseudomonas proteolytica, Pseudomonas meridiana, Arthrobacter gangotriensis, A. kerguelensis, Bacillus cecembensis* and *B. indicus* are some of the bacteria reported to successfully synthesize the silver nanoparticles *in vitro* (Iravani 2014; Panacek et al. 2006; Ahmad et al. 2003a, b; Mukherjee et al. 2001a). The microorganisms are known to have the capability to reduce the silver ions to silver nanoparticles which may be of different shapes depending upon the conditions of the culture and mechanism opted by the bacteria. Usually light is required for the biosynthesis of silver nanoparticles (Fayaz et al. 2010).

Klaus et al. (1999) investigated the bacteria from silver mines. This bacterium *Pseudomonas stutzeri* produce a large number of nanoparticles when cultured in the medium supplemented with silver nitrate. They observed the accumulation of silver nanoparticles particles in the periplasmic space of bacteria so easy to get them. Similar behavior was reported from *Aspergillus flavus, Fusarium oxysporum, Verticillium sp.* In which a film was formed at the cell surface (Vigneshwaran et al. Vigneshwaran et al. 2007; Bhainsa and D'Souze 2006) (Fig. 9.2).

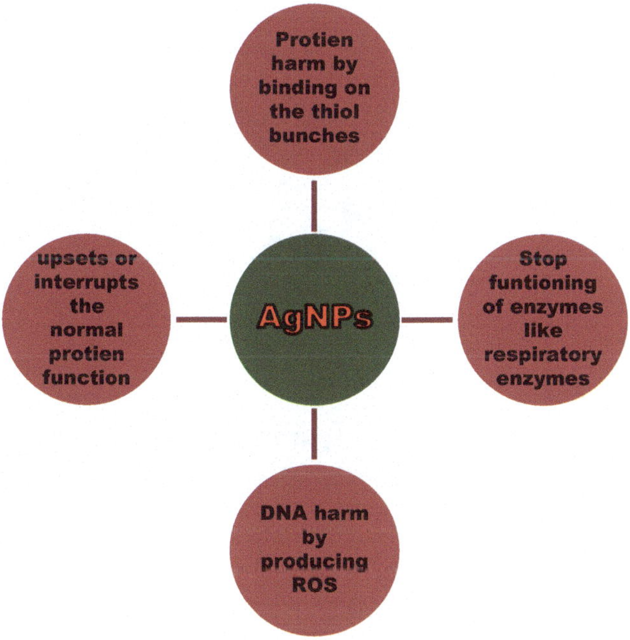

Fig. 9.2 Harmful effects of silver on bacteria

9.2.4 Alloy Nanoparticles

Alloy nanoparticles (ANPs) have gained much attention in the fields of optical materials electronics, catalysts, coatings and medicines. Point of interest here is that microbes are also involved in this complex process of bimetallic nanoparticles formation. *Fusarium oxisporum* is found to form AU-AG composite nanoparticles and their mechanism of synthesis is dependent upon role of cofactor NADH [30]. Yeast cells have also been reported to synthesize the AU-Ag alloy NPs (Senapati et al. 2005a). Fluorescence microscope and transmission electron microscope (TEM) studies showed that these Au-Ag alloy nanoparticles were of irregular shape (Senapati et al. 2005b) (Fig. 9.3).

9.2.5 Other Metallic Nanoparticles

Room temperature conditions (pH 8.0) advance the combination of uniform-sized 2–5 nm, round, and monodispersed intracellular mercury nanoparticles. The palladium nanoparticles are additionally revealed in the writing by the sulfate-reducing bacterium, *Desulfovibrio desulfuricans*, and metal particle decreasing bacterium, *Staphylococcus oneidensis* (Konishi et al. 2007; Lloyd et al. 1998).

Fig. 9.3 Nano alloy crystals

9.2.6 Oxide Nanoparticles

Microbes are also involved in the biosynthesis of metal oxide nanoparticles. Metal oxides may be of two types i.e.,

(i) Magnetic oxide nanoparticles
(ii) Nonmagnetic oxide nanoparticles

(i) **Magnetic Oxide Nanoparticles**

Magnetic nanoparticles like Fe_2O_3 (maghemite) and Fe_3O_4 (magnetite) are known to be produced naturally and can be used even in living systems. A lot of research work has been done on these types of particles because these are used in different applied fields. These particles are used in fields like treatments of different diseases, DNA examination, MRI imaging, Cancer treatments, Guided drug delivery and DNA examination (Windt et al. 2005).

Magnetotactic bacteria are present naturally and a very suitable example to explain the synthesis of magnetic nanoparticles in a natural way. These bacteria incorporate some iron oxides or sulfides or both, which are then known as the bacterial magnetic particles made up of Ferric oxide or Ferric sulphide. These are present inside the bacterial cell in chain manner and act as the natural bacteria compass needles to move along the oxygen concentrations (Fan et al. 2009; Bazylinski et al. 1994). These magnetic particles are nano-sized and enveloped by lipid bilayers (made of phospholipids). They are also fossilized.

Since 1975, when first magnetotactic microbes were reported, a number of bacterial strains including cocci, vibrios, spirilla, ovoid microorganisms, and multicellular microscopic organisms have been found to be magnetic in nature. These organisms occupy the aquatic environments while cocci forms prefer to inhabit the aquatic rocks' surfaces randomly (Fig. 9.4).

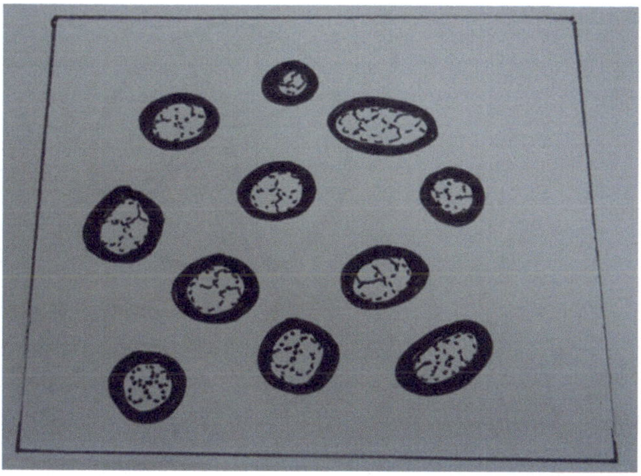

Fig. 9.4 TEM images (Sketches of magnetic NPs formed from *Bacillus sp.*)

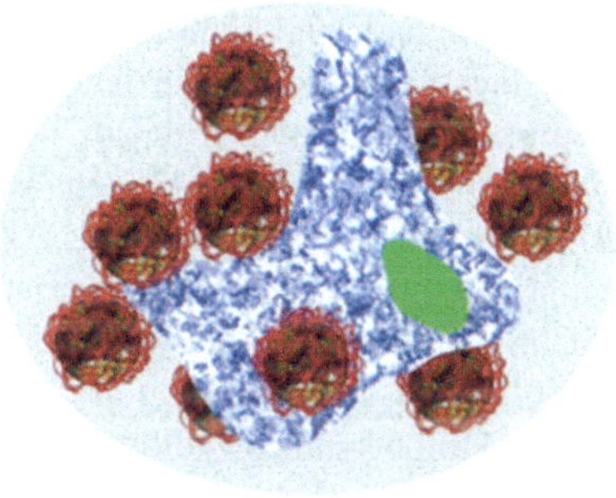

Fig. 9.5 Nonmagnetic oxide nanoparticles (Wu et al. 2008)

(ii) **Nonmagnetic Oxide Nanoparticles**

Metal oxide nanoparticles, other than the magnetic oxide nanoparticles have also been studied extensively. These include like zinc oxide nanoparticles, silicon oxide nanoparticles, Zirconium oxide nanoparticles, Silicon oxide and titanium oxide nanoparticles. These nanoparticles are playing their role in the fields of electronics, semiconductors, transistors, solar cells and even in medicines. Jha and co-workers found *Saccharomyces cerevisiae* mediated biosynthesis of Sb_2O_3 nanoparticles which is a green low-cost and reproducible method (Jha and Prasad 2010; Jha et al. 2009a, b; Bansal et al. 2005). Analysis indicated that Sb_2O_3 nanoparticles unit was a spherical aggregate having a size of 2–10 nm (Bai et al. 2006). Bansal and his coworkers used *Fusarium oxysporum* (Fungus) to produce SiO_2 and TiO_2 nanoparticles from aqueous anionic complexes of Si and Ti, respectively [44]. They also synthesized tetragonal $BaTiO_3$ and quasi-spherical ZrO_2 nanoparticles from *F. oxysporum* with a size range of 4–5 nm and 3–11 nm, respectively (Bansal et al. 2004, 2006) (Fig. 9.5).

Fig. 9.6 Sulphide nanoparticles

9.2.7 Sulfide Nanoparticles

Sulphide nanoparticles have their own importance in basic as well as specialized fields of research. Microorganisms are also playing their roles in this field and preparing sulphides.

Cunningham and Lundie (1993) found that *Clostridium thermoaceticum* promote the synthesis of CdS nanoparticles on the cell surface when grown in the medium supplemented with $CdCl_2$ with cysteine hydrochloride. *Klebsiella pneumonia* when provided with Cd^{2+} ions was able to convert them to CdS nanoparticles of 20–200 nm in size (Bansal et al. 2004). *E. coli* also plays its role when is supplied with $CaCl_2$ and Na_2SO_4 and gave rise to nanocrystals (Sweeney et al. 2004). Reports of CdS nanoparticles with yeast (*Candida glabrata*) are also present (Bai and Zhang 2009). Some other metal sulfide nanoparticles like ZnS, PbS, MoS have been reported to be synthesized by *F. oxysporium* extra-cellularly when react with metal salt at normal pH, Temperature and moisture conditions of the medium (Suresh et al. 2011; Bai et al. 2006) (Fig. 9.6).

9.3 Microbes and Inorganic World

As described earlier in the introduction that biological procedures are better and more effective approaches for the synthesis of nanoparticles. Chemical and physical methods are available and can create a large number of nanoparticles of specific shape and size but at the same time we face many odd situation; for example production and addition of hazardous by products, exposure of surroundings to strong chemicals and to strong radiations. Complicated machinery and procedures are required which adds to the cost of product. But when we opt some biogenic machinery like plants, bacteria or fungi, we find a green route which is environment friendly, consuming lesser or no chemicals with least requirement of the energy. These all facts make these green or biogenic systems as cost effective too (Gericke and Pinches 2006). Even the marine organisms have not been still exploited for the nanoparticles synthesis. But they can be used at a larger scale as Albeit, marine water is one of the wealthiest wellsprings of gold in the nature. There is no huge distribution with respect to usage of marine smaller scale living beings to deliver gold nanoparticles. Along these lines, there may be a plausibility of investigating marine microscopic organisms as nanofactories for AuNP biosynthesis (Mittal et al. 2013).

Since the life came into being, the living world and inorganic materials are in touch to produce different cycles to maintain life. Now science is much interested in digging out the truth about this organic and inorganic relationship. The chain of this knowledge continues when we see that a number of microbes can produce inorganic metal nanoparticles on their cell surface or in their environment (Jain et al. 2011; Kalishwaralal et al. 2010; Pugazhenthiran et al. 2009; Ahmad et al. 2003a, b, 2007).

Microorganisms provide a large surface area for contact of metal salts and enzymes to form nanoparticles. These methods of contact may be of two types i.e.,

(i) Intracellular synthesis
(ii) Extracellular synthesis

(i) **Intracellular Synthesis**

The synthesis of nanoparticles by microbes inside the cell is the intracellular synthesis. It is usually done with the help of enzymes and coenzymes in the cell. The nanoparticles are then released by treatment with detergents or by ultrasounds which is an additional step for final product separation. An example is the production of silver nanoparticles by *E. coli* and *Vibrio alginolyticus* (Rajeshkumar et al. 2013; Saifuddin et al. 2009).

(ii) **Extracellular Synthesis**

The precipitation of nanoparticles in the external environment of a cell by microbe is known as the extracellular synthesis of the nanoparticles. It is the area of interest at times for the nanomicrobiologist as it sounds to be cheap process with requirement of simple downstream processing. It also favors cost effective and large scale production of nanoparticles. A number of microbes are studied for their potential of extracellular synthesis of nanoparticles like *Pseudomonas proteolytica, Thermomonospora spp., Klebsiella pneumonia, Shewanella oneidensis, Pseudomonas meridiana, Arthrobacter kerguelensis, Bacillus indicus,* etc. (Velusamy et al. 2016; Arayanan et al. 2013). It has also been hypothesized that the organisms having resistance machinery to specific metal can only synthesize nanoparticles (Duran et al. 2005). Bacteria resist to heavy metal toxin by releasing toxic chemicals which act as the capping and reducing agents for the NPs synthesis. These chemicals include proteins, aminoacids, vitamins, enzymes, polysaccharides etc. Mostly involved component is Nitrate reductase (Deepak et al. 2011; Kumar et al. 2007).

Extracellular synthesis of nanoparticles is also done by the energy dependent ion efflux systems from their surface membrane proteins. These proteins may be some ATPase or cation-anion antitransporters. This method includes biosorption (cell walls of microbes have a number of points and groups where metal parts can bind), Biomineralization (the process in which bacteria release minerals on their surface to stiffen the surface) and complexation etc. (Castro et al. 2014).

9.4 Common Steps of NPs Synthesis by Microbes

Mostly the method opted includes following steps as such or sometimes with modifications:

- **Source of microorganism**

Initially the bacterial strains are arranged from some authentic source then they are maintained on bacterial Media slant at 27 °C as well as subcultured from time to time to regulate their viability

- **Production of biomass**

The strains are cultured, to produce the biomass for biosynthesis in different liquid broth media according to the nature of the strain used. In this method the cultured flasks are then incubated on an orbital shaker at 27 °C and agitated on orbital shaker. The biomass is harvested after 24 h and centrifuged at 12000 rpm for 10 min. The supernatant material is collected for the further reaction to synthesize nanoparticles.

- **Synthesis of silver nanoparticles**

Metal salts are then added to the culture as silver nitrate in this case. Amount and strength of salt depends upon the type of metal in use as well as on the metabolic machinery of the microbe used.

- **Characterization of silver nanoparticles**.

The bio reduction of the Ag + ions in the solution is usually visible due to color change and can be supervised easily. For further confirmation sample of 2 ml is withdrawn at altered time intervals and the absorbance is measured at a resolution of 1 nm using UV–visible spectrophotometer with samples in quartz cuvette.

- **Particle sizing measurements**

Particle size analyzing experiments are carried out by means of different techniques like SEM, EDX, particle size analyzer etc.

9.5 Nanoparticles Synthesis by Bacteria: Mechanism

Nanoparticles which are produced by the biological methods like by plants, fungi or bacteria are very effective. These biogenic methods are very helpful in a number of ways. Particularly when microbes are involved, then shifting of pH, temperature and medium composition is not a big deal and usually under control. Response time is very short due to their short life cycles. Research has focused microbes now for production of commercial scale amounts of nanoparticles as they are much abundant in the environment. They can even survive in the extreme conditions like extreme acidic and basic pHs, extreme temperature etc. they are even very modest in living. It is also evident from the literature that by just controlling the above mentioned factors like temperature, pH, minerals in the medium, water, oxygen availability, light, shaking etc., the shape and size of Metal NPs can be controlled. It is important to note that shape and size of nanoparticles is very essential for their practical applications in optics, drug delivery systems, and transistors etc. (Natarajan et al. 2010; Kalimathu et al. 2008; Duran et al. 2005).

Bacteria can produce NPs by reduction or precipitation of insoluble nontoxic metal Nano clusters. Mokhari et al. (2009) also suggested that photo-reduction mechanism of nitroreductase with some involvement of e^- donors and acceptors is responsible for metal nanoparticles synthesis. Kalimuthu et al. 2008 reported the synthesis of AgNPs by *Bacillus licheniformis* by nitrate reductase enzyme. This bacterial strain is well known to secrete a nitrogen reductase enzyme which is dependent upon a NADH co-factor (Fig. 9.7).

Konishi et al. (2006) reported molecular hydrogen as electron donor and involvement of hydrogenase enzyme which is responsible for the reduction of metal ions. Same activity of microbes was reported earlier by Woolfolk and Whiteley (1962) for U^{6+} reduction by *Micrococcus lactyliticus* (Fig. 9.8).

Sometimes, mechanisms become, multistep and more complex with the involvement of more than one component. Like pointed out that bioreduction of Pt NPs by sulfide reducing bacteria is a two step process.

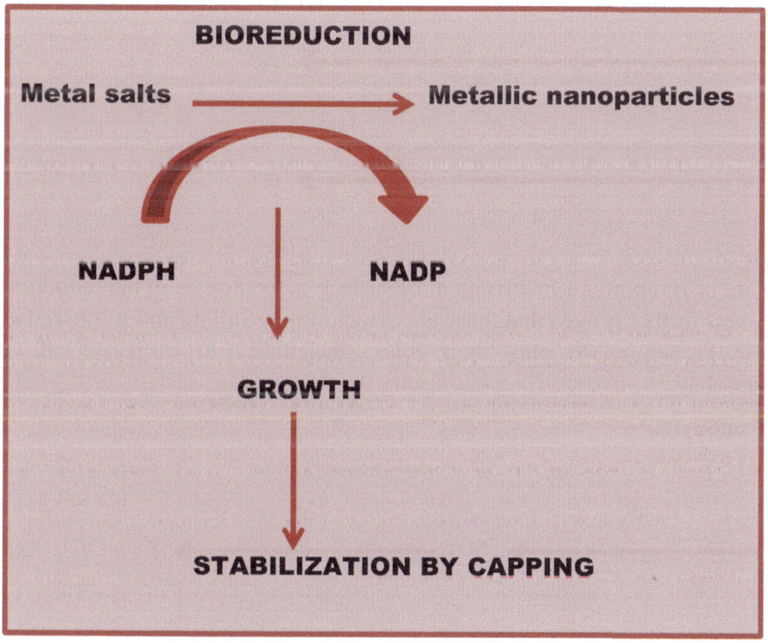

Fig. 9.7 Mechanism of NADH dependent mechanism

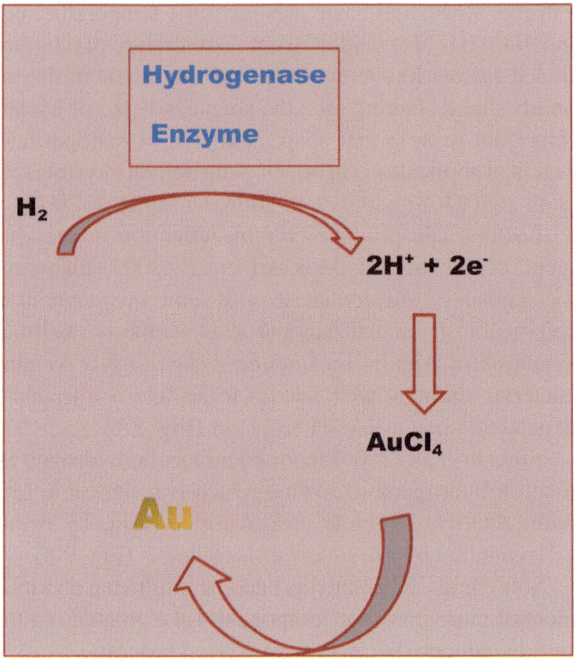

Fig. 9.8 Mechanism of Gold nanoparticles synthesis with Hydrogen as electron acceptor

a: oxygen sensitive cytoplasmic hydrogenase
b: Periplasmic hydrogenase

$$Pt^{+4} \xrightarrow{\quad a \quad} Pt^{+2}$$

$$Pt^{+2} \xrightarrow{\quad b \quad} Pt^{0}$$

There are also reports of involvement of surface cytochromes in the reduction process. It was further proved that these are cytochromes. Sulfate and sulfite reductases are also known as to play their roles particularly in formation of sulfide nanoparticles.

C: ATP sulfarylase
D: Sulfite reductase

$$SO_4 \xrightarrow{\quad c \quad} SO_3 \xrightarrow{\quad d \quad} S$$

Table 9.1 Some potential bacterial strains for nanoparticles synthesis

No. of Obs	Bacterial strain	Nanoparticles synthesis	Size (nm)	Intracellular/ Extracellular	References
1	*Bacillus subtilis*	Ag	5–60	Extracellular	Saifuddin et al. (2009)
2	*Bacillus cereus*	Au	4–5	Extracellular	Ganesh Babu and Gunasekaran (2009)
3	*E.Coli*	Pd,Pt	1–100	Extracellular	
4	Lactic acid bacteria	AsS	11.2	Extracellular	Sintubin et al. (2009)
6	*Proteus Mirabilis*	Au	10–20	Extracellular	Husseiny et al. (2007)
7	*Pseudomonas stutzeri*	Ag	200	Periplasmic	Thomas et al. (2012)
8	*Staphylococcus Aureus*	Au	1–100	Intracellular	Nanda and Saravanan (2009)

9.5.1 Parameters Involved

Mokhari et al. (2009) indicated that light is required for the improved and rapid synthesis of silver nanoparticles.

pH is another important parameter for almost all types of nanoparticles. Deplanche and Macaskie (2008) showed that at higher pH (pH = 6–7) the size of gold nanoparticles was higher as compared to the lower pH (pH = 2). Konishi et al. (2006) observed the microbial reduction of metal ions with lower pH. They observed that gold NPS of 10–20 nm in the periplasm of *Shewanella algae* cells at pH 7. While at pH 2.8 particles of 15–20 nm were produced on cell surface directly. At pH of 2, nanoparticles of 20 nm were produced intracellularly.

Holmes et al. (1997) found the role of bufferd medium in precipitation of CdS NPs by *Klebcella pneumonia.* Bari et al. presented cysteine desulfahydrase as the responsible enzyme for the synthesis of CdSNPs. At high and low pHs, protein are charged either positively or negatively (away from their isoelectric point). Low and high pH become functional for size and shape of nanoparticles as they are functional for proteins (Table 9.1).

9.6 Nanoparticles Synthesis Through Fungi

Fungi has been a good source for the synthesis of nanoparticles. *Geotrichum candidum* was the fungal strain used in 2013 for the production of gold nanoparticles with good size and shape with all their environment friendliness. The enzymes or proteins present in the microorganism might be responsible for the reduction of metal salts to nanoparticles. The AuNPs were characterized by UV-visible spectroscopy, dynamic light diffusing (DLS), vitality dispersive spectroscopy (EDS), examining electron magnifying instrument (SEM), and Fourier change infrared spectroscopy (FTIR). Nanoparticles were isolated by sonicating the entire cells after treatment with Tween 80. The entire cell intervened prepare demonstrated the oversimplified, achievable, simple to scale up, and minimal effort approach for the blend of AuNPs (Borse et al. 2015).

In the past platinum nanoparticles formation has also been ended by using cell free extracts of *Saccharomyces boulardii*. The formation was later on confirmed by UV-Vis spectroscopy. The surface images of platinum nanoparticles were taken by SEM. Platinum nanoparticles were visualized by transmission electron microscope (HRTEM-FEI) and Scanning electron Microscope (SEM) too (Magdi et al. 2014).

In the past the screening of nanoparticles have also be reported by using fungal sources as in 2015 some nanotechnologists used eight fungal strains namely, *Aspergillus fumigatus, Aspergillus flavus Candida albicans, Penicillium italicum, Syncephalastrum racemosum, Fusarium oxysporum Alternaria solani and Aspergillus ochraceus* for the screening of nanoparticles among which only six fungal species were found to reduce the silver salt into silver nanoparticles. The UV-visible spectra of the biosynthesized nanoparticles (AgNPs) by *Aspergillus ochraceus* (RCMB 036254) cell filtrate showed characteristic surface plasmon absorption at 420 nm. Transmission electron microscopy (TEM) micrograph showed the formation of spherical AgNPs ranging from 5.5 to 24.5 nm diameter. The elemental silver was characterized both qualitatively and quantitatively. The optimum conditions for maximum production of AgNPs were obtained using 0.75 mM silver nitrate at 40 °C and pH 6 (Park et al. 2011) (Fig. 9.9).

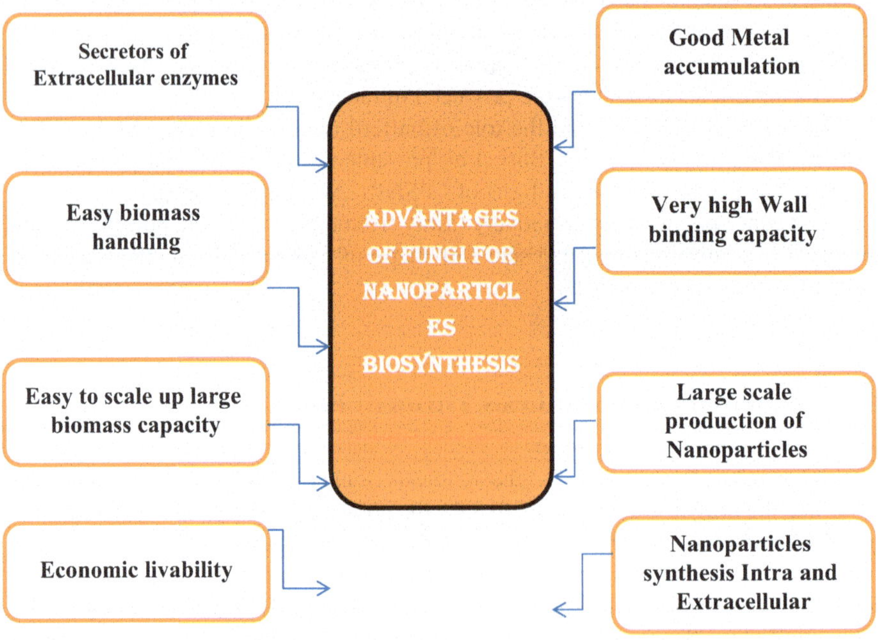

Fig. 9.9 Benefits of using Fungi as precursors of nanoparticles

Table 9.2 Some potential fungal strains for nanoparticles synthesis

No. of Obs.	Fungal strain	Nanoparticle synthesis	Shape	Application
1	*Aspergillus oryzae*	$FeCl_3$	Spherical	Agriculture, biomedical
2	*Aspergillus fumigatus*	ZnO	Sphecical Hexogonal	Industrial, medical, agriculture
3	*Rhizopus oryzae*	Au	Nanocrystalline	Pesticides
4	*Rhizopus stolonifer*	Au	Irregular	Pesticides
5	*Aspergillus niger*	Au	Polydispersed	Pesticides
6	*Fusarium oxysporum*	Au	Monodispersed	Optoelectronic
7	*Neurospora*	Au	Spherical	Anticancerous
8	*Verticillum sp.*	Au	Spherical, rod	Targeting cancer cells
9	*Sclerotoium sp.*	Au	Triangular, decahedral	Therapeutics
10	*Helminthosporium solani*	Au	Polydispersed	Anticancerous
11	*Trichoderma viridae*	Ag	Spherical	Agriculture, biomedical
12	*Volvariella volvaceae*	Au, Ag	Spherical	Agriculture, biomedical

Shobha and her coworkers (2014) synthesized Cu nanoparticles by using some fungal strains as *Penicillium aurantiogriseum, Penicillium citrinum* and *Penicilliumwaksmanii* in 2014. She also investigated the nanopartilces synthesis from algae *Bifurcaria Bifurcata*, the diterpenoids present in extract perform dual functions of reduction and stabilization of Cu Nps.

Trichoderma reesei also known as *Hypocrea jecorina*. AgNPs can be synthesized by this fungus, the fungus mycelium is exposed to the silver nitrate solution. That prompts the fungus to produce enzymes and metabolites for its own survival. In this process the toxic *Ag* + ions are reduced to the nontoxic metallic AgNPs through the catalytic effect of the extracellular enzyme and metabolites of the fungus. Fluorescence emission spectroscopy was used to produce detailed information on the progress of reduction of silver nitrate (formation of silver nanoparticles) on the nanosecond timescale. Fourier transform infrared spectroscopy was used for quantitative analysis of the reaction products (Ahmed et al. 2012) (Table 9.2).

9.7 Conclusion

Microbes are very minute organisms which are easy to handle and culture. They live all around us. Naturally they are beneficial as well as harmful. These microorganisms can be exploited easily for the production of nano-sized materials like nanoparticles, quantum dots, nanorods etc. It can be done by carefully selecting the type of microorganism, manipulating the culture conditions like nutrients, temperature, moisture content, pressure, pH etc. Nanoparticles of desired shape and size can be obtained. All the mechanism which is involved in nanoparticles synthesis is the reduction by the microbial enzymes with the help of some electron acceptors and electron donors varying from species to specie and type of metallic salt in the environment. Same species can produce different types of nanoparticles with different shapes and sizes. A lot of research work has been done with a number of microbes. Now the need of the hour is to optimize the each case for commercial scale production of nanoparticles of required size and shape.

References

Abou El-Nour MM, Eftaiha A, Al-Warthan A, Amma RAA (2010) Synthesis and application of silver nanoparticles. Arb J Chem 3(3):135–140

Ahmad A, Mukherjee P, Senapati S, Mandal D, Khan MI et al (2003a) Extracellular biosynthesis of silver nanoparticles using the fungus *Fusarium oxysporum*. Coll Surf B 28(02):313–318

Ahmad A, Senapati S, Khan MI (2003b) Intracellular synthesis of gold nanoparticles by a novel alkalotolerant actinomycete, *Rhodococcus* species. Nanotechnology 14(7):824–828

Ahmad R, Minaeian S, Shahverdi HR et al (2007) Rapid synthesis of silver nanoparticles using culture supernatants of Enterobacteria: a novel biological approach. Process Biochem 42:919–923

Ahmed AS, Oves M, Khan MS, Habib SS, Memic A (2012) Antimicrobial activity of metal oxide nanoparticles against gram-positive and gram-negative bacteria: a comparative study. Int J Nanomed 7:6003–6009

Arayanan PVB, Thangavelu D, Muthukumarasamy VK, Munusamy C, Gurunathan B (2013) Biological synthesis and characterization of intracellular gold nanoparticles using biomass of *Aspergillus fumigatus*. Bull Mater Sci 36(7):1201–1205

Azam AZ, Davood F, Ali RM, Muhammad N et al (2009) Synthesis and characterization of gold nanoparticles by tryptophane. Am J App Sci 82(4):691–695

Bai HJ, Zhang ZM (2009) Microbial synthesis of semiconductor lead sulfide nanoparticles using immobilized *Rhodobacter sphaeroides*. Mat Lett 63(9–10):764–766

Bai HJ, Zhang ZM, Gong J (2006) Biological synthesis of semiconductor zinc sulfide nanoparticles by immobilized *Rhodobacte rsphaeroides*. Biotechnol Lett 28(14):1135–1139

Bansal V, Rautaray D, Ahmad A, Sastry M (2004) Biosynthesis of zirconia nanoparticles using the fungus *Fusarium oxysporum*. J Mat Chem 14(22):3303–3305

Bansal V, Rautaray D, Bharde A (2005) Fungus-mediated biosynthesis of silica and titania particles. J Mat Chem 15(26):2583–2589

Bansal V, Poddar P, Ahmad A, Sastry M (2006) Room-temperature biosynthesis of ferroelectric barium titanate nanoparticles. J Am Chem Soc 128(36):11958–11963

Bazylinski DA, Garratt-Reed AJ, Frankel RB (1994) Electron microscopic studies of magnetosomes in magnetotactic bacteria. Microscopy Res Tech 27(5):389–401

Bhainsa KC, D'Souza SF (2006) Extracellular biosynthesis of silver nanoparticles using the fungus *Aspergillus fumigatus*. Coll Surf B 47(2):160–164

Borse V, Kaler A, Banerjee UC (2015) Microbial synthesis of platinum nanoparticles and evaluation of their anticancer activity. Int J Emerg Trends Elec Electron 11(2):65–73

Byrappa K, Ohara S, Adschiri T (2008) Nanoparticles synthesis using supercritical fluid technology-towards biomedical applications. Adv Drug Del Rev 60(3):299–327

Castro L, Blazquez ML, Munoz JA et al (2014) Mechanism and applications of metal nanoparticles prepared by bio-mediated process. Rev Adv Sci Eng 3:1–18

Cunningham DP, Lundie LL (1993) Precipitation of cadmium by *Clostridium thernoaceticum*. Appl Env Mic 59:7–14

Daniel MC, Astruc D (2004) Gold nanoparticles: assembly, supramolecular chemistry, quantum-size-related properties, and applications toward biology, catalysis, and nanotechnology. Chem Rev 104(1):293–346

Dean JA (1979) Lange's handbook of chemistry, 12th edn. McGraw-Hill Inc., New York, pp 6-2-6-19

Deepak V, Kalishwaralal K, Ram S K P, Gurunathan S (2011) M. Rai and N. Duran (eds) Metal nanoparticles in microbiology, Springer, Berlin/Heidelberg 2011, doi:https://doi.org/10.1007/978-3-642-18312-6_2

Duran N, Priscyla D, Marcato PD, Alves O, De Souza G, Esposito E (2005) Mechanistic aspects of biosynthesis of silver nanoparticles by several *Fusarium oxysporum* strains. J Nanobiotech 3:1–7

Fan TX, Chow SK, Zhang D (2009) Biomorphic mineralization: from biology to materials. Progress Mat Sci 54(5):542–659

Fayaz AM, Balaji K, Girilal M, Yadav R et al (2010) Biogenic synthesis of silver nanoparticles and their synergistic effect with antibiotics: a study against gram-positive and gram-negative bacteria. Nanomed Nanotech Biol Med 6(1):103–109

Ganesh Babu MM, Gunasekaran P (2009) Production and structural characterization of crystalline silver nanoparticles from *Bacillus cereus* isolate. Coll Surf B 74(1):191–195

Gericke M, Pinches A (2006) Biological synthesis of metal nanoparticles. Hydrometallurgy 83(1–4):132–140

Hayat MA (1989) Colloidal gold: principles, methods, and applications. Academic, San Diego

Holmes JD, Richardson DJ, Saed S, Evans-Gowing R, Russell DA, Sodeau JR (1997) Cadmium-specific formation of metal sulfide "Q-particle" by *Klebsiella pneumoniae*. Microbiology 143:2521–2530

Husseiny MI, El-Aziz MA, Badr Y, Mahmoud MA (2007) Biosynthesis of gold nanoparticles using *Pseudomonas aeruginosa*. Spectrochim Acta A 67(3–4):1003–1006

Iravani S (2014) Bacteria in nanoparticle synthesis: current status and future prospects. Hindwai Pub Corp Int Sch Res Not 18:1–19

Jain N, Bhargava A, Majumdar S, Tarafdar JC, Panwar J (2011) Extracellular biosynthesis and characterization of silver nanoparticles using *Aspergillus flavus* NJP08: a mechanism perspective. Nanoscale 3:635–641

Jha AK, Prasad K (2010) Ferroelectric $BaTiO_3$ nanoparticles: biosynthesis and characterization. Coll Surf B 75(1):330–334

Jha AK, Prasad K, Prasad K (2009a) A green low-cost biosynthesis of Sb_2O_3 nanoparticles. Biochem Eng J 43(3):303–306

Jha AK, Prasad K, Kulkarni AR (2009b) Synthesis of TiO_2 nanoparticles using microorganisms. Coll Surf B 71(2):226–229

Kalimuthu K, Babu RS, Venkataraman D, Mohd B, Gurunathan S (2008) Biosynthesis of silver nanocrystals by *Bacillus licheniformis*. Coll Surf B 65:150–153

Kalishwaralal K, Deepak V, Pandian SRK (2010) Biosynthesis of silver and gold nanoparticles using *Brevi bacteriumcasei*. Coll Surf B 77:257–262

Klaus T, Joerger R, Olsson E, Granqvist CG (1999) Silver-based crystalline nanoparticles, microbially fabricated. Proc Natl Acad Sci U S A 96(24):13611–13614

Konishi Y, Tsukiyama T, Ohno K, Saitoh N, Nomura T, Nagamine S (2006) Intracellular recovery of gold by microbial reduction of AuCl ions using the anaerobic bacterium *Shewanella algae*. Hydrometallurgy 81:24–29

Konishi Y, Ohno K, Saitoh N (2007) Bioreductive deposition of platinum nanoparticles on the bacterium *Shewanella algae*. J Biotech 128(3):648–653

Kumar BL, Gopal DVRS (2015) Effective role of indigenous microorganisms for sustainable environment. Biotech 5:867–876

Kumar AS, Abyaneh MK, Sulabha SWG, Ahmad A, Khan MI (2007) Nitrate reductase mediated synthesis of silver nanoparticles from $AgNO_3$. Biotechnol Lett 29:439–445

Lengke MF, Fleet ME, Southam G (2006a) Morphology of gold nanoparticles synthesized by filamentous cyanobacteria from gold(I)-thiosulfate and gold(III)-chloride complexes. Langmuir 22(6):2780–2787

Lengke MF, Ravel B, Fleet BE et al (2006b) Mechanisms of gold bioaccumulation by filamentous cyanobacteria from gold(III)-chloride complex. Env Sci Tech 40(20):6304–6309

Liu J, Qiao SZ, Hu QH, Lu GQ (2011) Magnetic nanocomposites with mesoporous structures: synthesis and applications. Small 7(4):425–443

Lloyd JR, Yong P, Macaskie LE (1998) Enzymatic recovery of elemental palladium by using sulfate-reducing bacteria. App Env Micro 64(11):4607–4609

Luechinger NA, Grass RN, Athanassiou EK, Stark WJ (2010) Bottom-up fabrication of metal/metal nanocomposites from nanoparticles of immiscible metals. Chem Mat 22(1):155–160

Magdi HM, Mourad MHE, AbdelAziz MM (2014) Biosynthesis of silver nanoparticles using fungi and biological evaluation of mycosynthesized silver nanoparticles. Egypt J Exp Biol 10(1):1–12

Mittal AK, Kaler A, Mulay AV, Bannergee UC (2013) Synthesis of gold nanoparticles using whole cells of *Geotrichum candidum*. J Nanopart 1(1):1–7

Mokhari N, Daneshpajouh S, Seedbagheri S, Atashdehghan R, Abdi K, Sarkar S, Minaian S, Shahverdi HR, Shahverdi AR (2009) Biological synthesis of very small nanoparticles by culture supernatant of *Klebsiella pneumoniae*: the effects of visible-light irradiation and the liquid mixing process. Mater Res Bull 44:1415–1421

Mukherjee P, Ahmad A, Mandal D (2001a) Bioreduction of $AuCl4^-$ ions by the fungus, *Verticillium* sp. and surface trapping of the gold nanoparticles formed. Angew and teChemie Int Ed 40(19):3585–3588

Mukherjee P, Ahmad A, Mandal D (2001b) Fungus-mediated synthesis of silver nanoparticles and their immobilization in the mycelial matrix: a novel biological approach to nanoparticle synthesis. Nano Lett 1(10):515–519

Nanda A, Saravanan S (2009) Biosynthesis of silver nanoparticles from *Staphylococcus aureus* and its antimicrobial activity against MRSA and MRSE. Nanomedicine 5(4):452–456

Natarajan K, Elvaraj SS, Murty VR (2010) Microbial production of silver nanoparticle. Digest J Nanomat Biostr 5(1):135–140

Panáček A, Kvítek L, Prucek R (2006) Silver colloid nanoparticles: synthesis, characterization, and their antibacterial activity. J Phy Chem B 110(33):16248–16253

Park Y, Hong YN, Weyers A, Kim YS, Linhardt RJ (2011) Polysaccharides and phytochemicals: a natural reservoir for the green synthesis of gold and silver nanoparticles. IET Nanobiotech 5:69–78

Pugazhenthiran N, Anandan S, Kathiravan G et al (2009) Microbial synthesis of silver nanoparticles by *Bacillus* sp. J Nanopart Res 11:1811–1815

Rajeshkumar S, Malarkodi C, Paulkumar K et al (2013) Intracellular and extracellular biosynthesis of silver nanoparticles by using marine bacteria *Vibrio alginolyticus*. Nanosci Nanotechnol 3(1):21–25

Riddin T, Gericke M, Whiteley CG (2010) Biological synthesis of platinum nanoparticles: effect of initial metal concentration. Enz Mic Tech 46:501–505

Saifuddin N, Wong CW, NurYasumira AA (2009) Rapid biosynthesis of silver nanoparticles using culture supernatant of bacteria with microwave irradiation. J Chem 6:61–70

Senapati S, Mandal D, Ahmad A (2005a) Fungus mediated synthesis of silver nanoparticles: a novel biological approach. Ind J Phys A 78(1):101–105

Senapati S, Ahmad A, Khan MI et al (2005b) Extracellular biosynthesis of bimetallic Au-Ag alloy nanoparticles particles. Small 1(5):517–520

Shobha G, Moses V, Ananda S (2014) Biological synthesis of copper nanoparticles and its impact. Int J Pharm Sci Inven 3(8):28–38

Sintubin L, De Windt W, Dick J, Mast J, Ha DV, Verstraete W, Boon N (2009) Lactic acid bacteria as reducing and capping agent for the fast and efficient production of silver nanoparticles. Appl Microbiol Biotechnol 84(4):741–749

Stark AL (2010) Beneficial microorganisms: countering Microbephobia. CBE Life Sci Edu 9:387–389

Suresh AK, Pelletier DA, Wang W (2011) Biofabrication of discrete spherical gold nanoparticles using the metal-reducing bacterium *Shewanella oneidensis*. Acta Biomater 7(5):2148–2152

Sweeney RY, Mao C, Gao X (2004) Bacterial biosynthesis of cadmium sulfide nanocrystals. Chem Biol 11(11):1553–1559

Tan Y, Dai X, Li Y, Zhu D (2003) Preparation of gold, platinum, palladium and silver nanoparticles by the reduction of their salts with a weak reductant-potassium bitartrate. J Mater Chem 13:1069–1075

Thomas R, Viswan A, Mathew J, Radhakrishnan EK (2012) Evaluation of antibacterial activity of silver nanoparticles synthesized by a novel strain of marine *Pseudomonas sp*. Nano Biomed Eng 4:139–143

Velusamy P, Kumar GV, Jeyanthi V et al (2016) Inspired green nanoparticles: synthesis, mechanism, and antibacterial application. Toxicol Res 32(2):95–102

Vigneshwaran N, Ashtaputre NM, Varadarajan PV, Nachane RP et al (2007) Biological synthesis of silver nanoparticles using the fungus *Aspergillus flavus*. Mat Lett 61(6):1413–1418

Windt W, Aelterman P, Verstraete W (2005) Bioreductive deposition of palladium (0) nanoparticles on *Shewanella oneidensis* with catalytic activity towards reductive dechlorination of polychlorinated biphenyls. Environ Microbiol 7(3):314–325

Woolfolk CA, Whiteley HR (1962) Reduction of inorganic compounds with molecular hydrogen by *Micrococcus lactilyticus*. I. Stoichiometry with compounds of arsenic, selenium, tellurium, transition and other elements. J Bacterial 84:647–658

Wu W, He Q, Jiang C (2008) Magnetic iron oxide nanoparticles: synthesis and surface functionalization. Nanoscale Res Lett 3(11):397–415

Zhang X, Yan S, Tyagi RD, Surampalli RY (2011) Synthesis of nanoparticles by microorganisms and their application in enhancing microbiological reaction rates. Chemosphere 82(4):489–494

Chapter 10
Analytical Techniques in Nano Botany

Aneeqa Sabah Nazir

10.1 Analytical Techniques

Biomolecular nanoscience is a novel way of looking at the world of living system either plants or animals as well as microbes. It should be a productive attempt to get knowledge about life and biology. There is always a need to sort out, categorize and catalogue the detail of things and components of our whole world at macro, micro and of course the nano scale. Characterization and analysis is the means of communication for scientific achievements. Variety of machines, equipment, instruments, tools and computer hardware, software is required to accompany the different measurements. There are number of techniques like optical probe, electron probe, spectroscopic methods, scanning electron, transmission electron etc. to measures and analyze the things all around us. Most of these methods evolved for analysis in today's nano-field.

This chapter will elaborate the detail of analytical techniques and measuring parameters in the field of botany at nano-scale to acquire meaningful perspective.

Basically, the size and shape of nanoparticles provide an effective control and impressive changes in physico-chemical properties, so the synthesis, analysis and characterization of micro- and nanoparticles is becoming crucial in today's technical era. Furthermore, the functional micro- and nanomaterials based devices with unique and enhanced performance in the field of electronics, optics, photonics, spintronic, magnetism and ceramics etc. are in demand.

A. S. Nazir (✉)
Department of Physics, Lahore College for Women University, Lahore, Pakistan

S. Javad, A. Butt (eds.), *Nanobotany*, https://doi.org/10.1007/978-3-319-77119-9_10

10.1.1 Background

A key to all analysis is the light and matter interaction as depicted by scientist and researchers over several million years. Afterward, atoms and molecules and then an electron. The word spectroscopy originates from Latin: Spectrum, meaning: An appearance, scopy from Greek *skopein*," to view". It's about 2430 years ago, that first time lenses were introduced by Aristotle and further light must be needed to visualize the colors. The focusing properties of lenses were presented by Euclid. The phenomena of reflection, scattering, interference and refraction were added by Scientists such as Seneca, Cleomedes, and Ptolemy. The spectroscopic history is filled with many milestones but we will focus on electrons based probe analytical techniques and characterization methods.

Microscopy comes from ancient Greek word: Mikros, "small" and Skopeo to "examine" Microscopy means invisible to eye, the technique to observe the objects by the aid of an instrument.

There are three well known branches of microscopy: optical, electron and scanning probe. In electron microscopes, electrons are used as primary illuminating source instead of visible light in optical microscopes. High vacuum is required to start the operation and electromagnetic lenses are used to view the samples for analysis in an electron microscope.

The term electron was first introduced by G.J. Stoney in 1874, (Greek, electron: meaning "amber", a substance produces spark on rubbing). Electron is negatively charged (1.6022×10^{-19} C), stable and sub-atomic particle with rest mass m_e, equal to 9.1094×10^{-31} kg. In 1895, Jean Perrin determined that cathode rays also carry the negative charge. In 1877, J.J. Thomson discovered electrons corpuscular (e.g. particulate) nature as he found that electron trajectory is deflected by applying magnetic field. Charge to mass ratio (m/e) of an electron was determined by measuring the degree of deflection.

$$-e / m_e = -1.758820150(44) \times 10^{11} \mathrm{C.Kg}^{-1}$$

A famous oil-drop experiment was performed to measure Electron's charge by R.A. Millikan by calculating the total charge q on an oil drop of pre-determined mass $m_{drop.}$

$$q = m_{drop} g / E$$

Where g is gravitational constant

E is strength of applied magnetic field. By varying the magnetic field strength, multiple of 1.6×10^{-19} C was observed.

Wave-like properties of electron were proposed in 1924 by Louis de-Broglie and awarded Nobel prize in 1929. He theorized wavelength of a particle depends on the momentum.

10.1.2 The Nano Botany Perspective

Nanomaterials exhibit distinguishing properties like high surface area and reactivity, tunable size and morphology etc. Plenty of information available for physiological, bio-chemical and molecular mechanism of nanoparticles in plants for better plant growth and development (Manzer et al. 2015). The beauty and interesting properties of natural nano entities arouse curiosity for investigations. Increasing use plant extract for synthesis of nanoparticles in the field of botany for the growth, development in plant cell, it is necessary to analyze the effect of nano- plant composites and detect the path followed by nano-particles.

Characterization and analysis of nano scale objects is not possible without tools and techniques. Beside the measurement of macroscopic properties of certain material for example in powder form, we are now able to measure the properties of a single atom, a single molecule, a single cluster, a single colloidal particle, and even a single nanotube. Even now, the probe materials, not much larger than atoms or molecules can be observed and measured to develop integrated techniques. These analytical techniques can get information simultaneously, quickly, efficiently, and precisely. Extraction of molecular and atomic information from bulk response to stimulation is also applicable for nanomaterial analysis. However, the analysis of individual molecule and an atom is focusing point now. Over the past few years, characterization and analytical methods have undergone dramatic changes without any limitations to probe the secrets of individual Nano entities. The probes themselves of the order of the nanoscale, for example the visible light of hundreds of nanometers, x-rays on the order of nanometers, electron beams with sub nanometer wavelength, high-energy particles smaller than nanometers, and much more.

Natural nano materials exhibit variety of phenomena like self-cleaning by lotus leave due to hydrophobicity that can be observed by measuring corresponding contact angle. This is definitely an important field with applications in advanced surfaces and agriculture, Chlorophyll fluorimetery measuring photosynthesis efficiency and impact of nanoparticles on plant health, TEM and SEM to track nanoparticles in plants and plant cells, UV-Visible spectroscopy for band gap tuning, Atomic force microscopy AFM for inner structural imaging, measuring the surface topography at sub nanometer resolution and Raman spectroscopy to investigate cell structures, interaction of drugs with cells, diagnose of skin diseases etc. will be discussed in detail. Advanced separation techniques using nano porous materials coupled to GC-MS (gas chromatography and mass spectroscopy) to identify the different substances, surface enhanced Raman spectroscopy (SERS) to detect pesticides and plant chemicals coupled with computer based principal component analysis (PCA) for product identification like diamond for structure analysis, retrieving for smoothing the obtained results etc.

As nano is an interdisciplinary approach so all characterization techniques for nanoparticles and structures are merged for analysis SEM, TEM, Raman, STM, AFM, UV-visible and XRD etc. The versatility and power of these nanoscale tools have increased over the past few decades. The control, manipulation, and integration of individual atom has at last been accomplished due to advent of these methods.

One of the greatest developments in the twentieth century is the application of computer capability to analytical tools. In addition to their value in crunching data, computers also assist researchers with positioning samples, focusing images, monitoring reactions, and data acquisition and processing for analysis.

10.1.3 Types of Analytical Techniques

Nano technology is multidisciplinary approach in nature. Fairly, it is necessary to apply equally broad inventions of characterization and analytical techniques to all artificially prepare (synthetic) and naturally existing nano structurer and materials. There are approximately 700 single- signal and approximately 100 multisignal analytical techniques (Hornyak et al. 2008) based on combination of three kinds of physical phenomena, First *Primary analytical probes* (photons, electrons, neutrons, ions, etc.) including the input stresses (heat, pressure, temperature, electric and magnetic field, mechanical stress etc.), second, the measurable secondary effects (secondary electrons, stress and strain, mechanical deformation, heat and pressure, etc.), monitoring of measured data (time, temperature, energy, angle and phase) (Kelsall et al. 2005). Primary probe (1° Probe) modified into secondary (2°), the measuring signals by interaction with the matter. Excited electron and photon, Plasmon, phonons and excitons are some example of matter alterations (Table 10.1).

10.2 Optical Probe Techniques

Optical microscopic technique is primary tool for morphological analysis, optical properties and spatial orientation in the crystals at micro-level by polarizing the visible light. Visible light within the wavelength range 400 nm to 800 nm is being used as primary probe in optical characterization techniques and secondary effect is the photon arises from phenomena like elastic and inelastic scattering and emission etc. Optical methods are useful for imaging as well as chemical analysis as listed in Table 10.2. Biological tissue can also be analyzed with and without staining by using fluorescent dyes. Before going into the detail of analysis, it is necessary to know what actually is going on inside an optical microscope.

The relationship between focal length f, the distance from center of the lens to a focal point and the object and image distances is $1/f = 1/o + 1/i$. where o and i are the object and image distances. Magnification, M, is simply the ratio of image and object distances $M = i/o$.

Resolution comes from the Latin word resolutionem, "the process of reducing things into simpler forms"). It is the ability to separate something into its smaller components. Resolution gives the minimum distance between two distinguishable

Table 10.1 Types of analytical techniques for characterization at nano scale

Probe type	Technique	Analytical effect
Optical	1: Binocular microscopes Compound microscopes	1: Imaging/fine and gross morphology
	2: Confocal laser -scanning microscopy (CLSM) Scanning near- field optical microscopy (SNOM) Two photon fluorescence microscopy (2PFM)	2: Imaging /morphology Image screening Fluorophores/ biological samples
	3: Dynamic light scattering (DLS) Photon correlation spectroscopy (PCS) Quasi elastic spectroscopy (QELS)	3: Particle sizing
Electron	1: Scanning electron microscopy (SEM) and electron probe microanalysis(EFMA)	1: Raster imaging, topology, morphology, particle size and chemical analysis.
	2: Transmission electron microscopy (TEM) High- resolution transmission electron microscopy (HRTEM) Scanning transmission electron microscope (STEM)	2: Particle imaging, size and shapes Structural imaging and chemical analysis Biological samples
	3: Field- emission microscopy (FEM)	3: Surface structure and molecular properties
	4: Reflection high energy electron diffraction (HEER) Low-energy electron diffraction (LEED) Electron back scattered diffraction (EBSD) in SEM	4: Surface structure Surface and adsorbate bonding: Crystallographic information
	5: Electron energy loss spectroscopy (EELS)	5: Inelastic electron interaction
	6: Auger electron spectroscopy (AES)	6: Chemical surface analysis
Scanning	1: Atomic force microscopy (AFM)	1: Imaging, topology, surface structure
	2: Scanning tunneling microscopy (STM)	2: Surface structure, imaging and topology
	3: Atomic probe microscopy (APM) Field ion microscopy (FIM) Imaging atomic probe (IAP) Atomic probe tomography (ATP) Position-sensitive atomic probe (POSAP)	3: 3-D imaging Chemical investigation and atomic spacing Emitted ion from imaging surfaces Position-sensitive lateral location of atoms Mass position resolution

(continued)

Table 10.1 (continued)

Probe type	Technique	Analytical effect
Spectroscopic (photon)	1: Ultraviolet photoemission spectroscopy (UPS) UV-visible spectroscopy	1: Surface analysis Chemical analysis
	2: Atomic absorption spectroscopy(AAS) Atomic emission spectroscopy (AES) Fluorescence spectroscopy (FS) Inductively coupled plasma Spectroscopy (ICP)	2: Elemental analysis
	3: Surface plasmon resonance spectroscopy (SPR) Localized surface plasmon spectroscopy (LSPR) Photoluminescence spectroscopy (PLS)	3: Surface/ adsorbate analysis Nanoparticle analysis Elemental analysis
	4: Raman spectroscopy (RS) Surface- enhanced Raman spectroscopy (SERS) Single molecule detection (SERS) Fourier transform infrared spectroscopy (FT-IR) Near infra-red spectroscopy (NIR) Diffuse reflectance FTIR	4: Vibrational analysis Bond structure, chemical analysis Ability to probe single molecule Asymmetrical vibrational analysis Surface IR analysis Surface adsorbate analysis
	5: X-ray diffraction (XRD)	5: Crystal structure
	6: X-ray fluorescence (XRF) Energy dispersive x-ray spectroscopy (EDX) Wavelength dispersive x-ray spectroscopy (WDS)	6: Elemental analysis
	7: X-ray photoelectron spectroscopy (XPS) Small angle x-ray scattering (SAXS)	7: Surface analysis, depth profiling Particle size and surface analysis

Ion-particle	1: Mass spectroscopy (MS)	1: Composition of material
	2: Secondary ion mass spectroscopy (SIMS)	2: Composition of solid surfaces
	3: Rutherford back scattering (RBS)	3: Elemental analysis (qualitative and quantitative)
Thermodynamics	1: Thermal gravimetric analysis (TGA) Differential thermal analysis (DTA) Differential scanning calorimetry (DSC) Noncalorimetry (NC)	1: Mass loss against temperature Reaction heat, heat capacity Reaction heat and phase changes Latent heat of fusion
	2: Temperature programmed desorption (TPD) Single crystal absorption calorimetry (SCAC) Linear-volume thermodilatometry (TDM)	2: Surface adsorbate properties Adsorption and adhesion energies Dimensions as function of T
	3: Thermoluminescence (TL)	3: Surface states detrapping
	4: Brunauer-Emmett-teller (BET) Mercury porosimetry (MP) Barnett–Joyner–Halenda method (BJH) SEARS method (SEARS)	4: Surface area analysis Pore volume, pore size Pore size distribution Colloidal size, specific surface area

Table 10.2 Advantages and disadvantages of microscopic techniques for biological samples

Microscopic technique	Sample states	Advantages	Disadvantages
Bright-field	Live or stained cells	Simple preparation	Biological samples need to be stained due to low contrast and natural pigmentation. Resolution is restricted to ~0.2 μm as staining creates the artifacts.
Dark field	Live or unstained cells	Simple set up Provide contrast to unstained cells and tissues	Cells need to be illuminated strongly, soft part may be damaged
Phase contrast	Transparent, unstained live cells	Fine details can be observed in high contrast and invisible under bright-field optics.	Thick samples appear to be distorted phase artifacts may be present
Fluorescence microscopy	To visualize the fluorescencepattern instained tissues.	Bio-molecules can be fluorescently stained. Immunofluorescence cells can be modified genetically to trace out the location of protein molecule through fluorescent signal in living organisms.	The fluorescence fades out due to photobleaching. Specimens with antibodies need to be fixed either chemically or treated with detergents to permeabilize cell membranes. Artifacts created by both procedures.
1: Super resolution fluorescence microscopy 2: Structured illumination microscopy (SIM) 3: Stimulated emission depletion microscopy (STED) 4: Optical reconstruction microscopy (STORM) 5: Photo-activated localization microscopy (PALM)	Resolve microscopic images by the factor of two over light diffraction	Diffraction resolution is ~200–250 nm SIM resolution ~ 100 nm STED resolution is approx. Several folds 0~70 nm. STORM: Resolution improved within the range ~ 10–55 nm PALM: Resolution improved from 10–55 nm, can track a singles molecule within the cell.	Effective image resolution by the 1: Intrinsic optical resolution 2: The labeling density 3: The size of the fluorescent materials.

TEM	Study of intercellular details of cells and tissues	Intracellular details can be visualized to a resolution of 0.2 nm Proteins or other molecules can be located on thin sections by applying antibodies attached to gold particles	Sample preparation techniques are lengthy and complex e.g. fixation (with chemicals or cryo techniques) introduce artifactually damage
TEM tomography	A branch of TEM with 3D reconstruction of a cell from a series of tilted 2D images.	3D images are formed from a thin section at 5–10-nm resolution.	TEM only make it possible for the sample to be tilted ±60/70°, so not a true 3D image
Cryo TEM	Samples are viewed in a vitreous state	Specimens are observed close to their original state without any application of chemicals to alter the native structures. This technique can be coupled with tomography.	Produces low-contrast images for unstained samples. Expensive and highly sophisticated machinery required. Samples sectioning in frozen state is very complex and require technical handling.
Cryo-EM-single particle analysis	To visualize macromoleculesand their complexes at sub-nanometer scale.	Applied to a homogeneous sample; e.g. protein complexes, and viruses etc. The series of digital images are obtained and used to reconstruct 3D perspectives of the particles of interest at near atomic resolution without any need of X-ray crystallography	Molecules/and their complexes cannot be crystallized.
SEM	To view the surface analysis of samples at resolutions of less than 1 nm	Images have a good depth of field that is useful for viewing biological samples and their surface morphology. Many techniques like immunogold labeling can be combined with SEM, for detection of macro-molecules and lipid membranes.	Cellular features of biological samples are destroyed due to dehydration as required for SEM analysis. Conductive coating on the sample obscure the important cellular details.

objects from set minimum distance. Resolution can be calculated by counting the pixels as well. Ability to measure the angular separation of closer images is called resolving power. The shorter the wavelength of the source is, the greater is the resolving power of the instrument, thus sample imaging is dependent on the source wavelength either optical or electronic analysis.

The resolving ability of optical microscopes is 0.35 μm. Resolution can be improved to 0.18 μm, by applying an oil emulsion to an objective lens. The resolution is improved to 0.18 μm (180 nm). Resolution of an optical microscope is generally equal to approximately half the wavelength of incident beam, the visible light. The wavelength of visible light lies roughly between 400 and 700 nm. Resolving power of electron microscope depends on applied accelerating voltage ~ 0.2 nm. Hardware quality and operational setting also affect the resolving power of instruments. Numerical aperture (N.A) of an optical system is defined as N.A = nsinθ, where n is refractive index of the medium between lens and sample, θ is the half angle subtended by the incident light rays entering the objective lens. Numerical aperture is dimensionless quantity represents the angular range of lens system to absorb and emit the visible light. The resolving power of the lens is proportional to wavelength divided by the numerical aperture R = λ/N.A. The resolution can be improved by (1) shortening the wavelength of the impinging radiation (e.g., the λ of blue light is shorter than λ of red light), (2) increasing value of refractive index e.g., air with n = 1.0 vs. oil with n ~ 1.5 and (3) increasing the N.A. Larger the half-angle θ, bigger the N.A). N.A. increases as focal length f decreases (Fig. 10.1a).

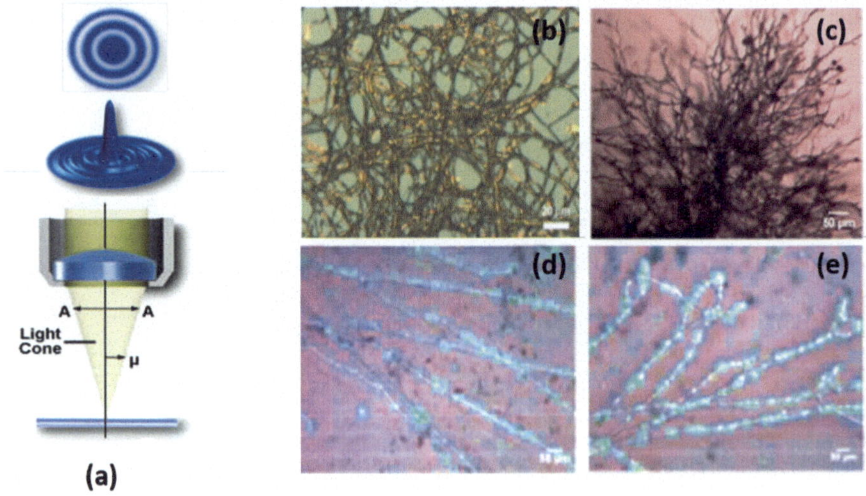

Fig 10.1 (**a**) Lens configuration highlighting the numerical aperture and corresponding air discs of microscope. When light passes through an aperture, specimen appears to show small concentric circles in an image, called an airy disc. Airy discs are formed by the diffraction of light passing through the circular aperture of objective lens. (**b, c, d, e**) Optical micrographs for gold-templated microwires at different resolutions. (**d, e**) Brightness of wires exhibited reflection of gold nanoparticles physisorbed on fungal hyphae

In optical microscope, the Airy projection is made of concentric rings with increasing thickness and decreasing spacing between the rings from the center. Diameter of central Airy disk is inversely proportional to the diameter of the aperture. Each object in the image is diffracting the impinging electron or an optical beam, so able to form its own Airy disk. But unfocused and unresolved images formed confused airy disks. The Abbe diffraction barrier limits resolution. Minimum resolving of an object can be measured by Rayleigh criteria. The Rayleigh criterion states that when the central maximum of the first image falls into the first minimum of the diffraction pattern of the second image, two such images are considered to be just resolved. For light transiting through a slit with width equal to d, the Rayleigh criterion is

$$d\sin\theta_{R=}\ \lambda;\sin\theta_{R=\lambda/d};\theta R=\lambda/d$$

For small angles: sin θR ~ θR. Therefore, the Rayleigh criterion minimum angular resolution for circular aperture of diameter D can be given as follows:

$$\text{Angular Resolution}=\theta_R=1.22\lambda/2\,N.A=1.22\lambda/d=0.61\lambda/N.A$$

where D is the diameter of the circular aperture. For analyzing the materials at nanoscale, angular resolution is converted into spatial resolution by multiplying by the focal length f. If the distance R is the arc subtended by θmin, then R = f θ_R and

$$\text{Spatial resolution}=f\,\theta_R=1.22\lambda f/D$$

Depth of focus D_{focus} is defined as distance above and below the focal plane, where an object is clearly focused. Depth of focus is influenced by an aperture size, an angle of divergence (ratio of aperture radius and working distance α, tan α ~ α) and focal length of the objective lens. The depth of focus for an electron microscope is about 300 times that of an optical microscope.

$$\tan\alpha=R_0/W\sim\alpha.\text{Depth of Focus}=D_{focus}=0.2/\alpha M\mu m,$$

where α is an angle formed between aperture and focal point of sample. Where M is the magnification. D_{focus} may also be described as ratio of wavelength to the square of N.A. as D_{focus} ~λ/(N.A).

Increase in D_{focus} by reducing the aperture diameter results in a decreased signal to-noise ratio (Goldstein et al. 1992). Depth of field (D_{field}) is different from depth of focus. Depth of field defines the lateral range focused and is described mathematically as D_{field} = 0.61 λ / nsinαtanα, where α is the semi-angle subtended by the aperture as before (Tay et al. 2013).

Mathematics of magnification, resolution, and diffraction for both optical and electron microscope are essentially the same. Confocal microscope was invented by Marvin Minsky in 1953. In confocal microscope, images are illuminated through a circular aperture and the fluorescent signal is channeled through another pinhole near the detector. Finally, images are collected by laser scanning, even the images that are out of focus are scanned easily. In near-field microscopes, a glass fiber of few nanometers above the sample surface produces optical images in the order of 50-nm resolution. Image is scanned by scanning near-field optical microscopy (SNOM) provides quite a breakthrough in optical imaging. Unwanted reflections and emissions are blocked by confocal microscopy. Confocal microscopes usually operate under the fluorescent mode but also absorb some part of reflected light to produce the images. The resolution of confocal microscope ~350 nm is better than that of simple optical microscopes ~130 nm respectively. Contrast is defined as the difference in intensity between an object, background and other nearby objects. Quality of both digital and analogue images depends on three main factors: 1: specimen preparation technique, 2: optical imaging, 3: recorded and analyzing system for data interpretation. Chromatic aberration may occur by using the polychromatic beam and the lens imperfections as focal length is dependent on wavelength of incident light. When the incident radiation is out of focus due to some lens imperfections, spherical aberrations arise. Spherical aberration can be resolved by adjusting an aperture size. Unequal focusing of vertical and horizontal lines at different points along optical axis forms the images at alternative points, called astigmatism, another aberration by light microscope.

In order to study all analytical techniques, it is necessary to understand some typical optical terms. Interference is defined as the superposition of electromagnetic waves, as electron waves and particle interfere either constructively or destructively depending on type of material and structure. The colors observed in thin films by soap bubbles display an interference phenomenon. Diffraction is the bending of waves when it encounters an obstacle. Airy disks in optical microscope are formed by diffraction of point objects. Reflection of light occurs when an incoming wave is symmetrically bounced back from the surface. Refraction is the deviation of light wave, when travel from one material to another. Absorption and emission occur by interaction of electrons of a material interact with EM radiation— the excitation and relaxation of electronic energy states. Transmission and reflection of light waves occur due to difference in frequencies of incident light and natural frequencies of vibrating materials. Transmission of radiation through a material occurs after absorbing and reflecting some part of material. Scattering is the deflection of light rays from the original straight path and scattered in many directions. Most commonly observed from the surfaces of particle having size larger than 20 to 50 nm but smaller than bulk entities.

Optical microscopy is useful in the field of medical sciences and engineering. Thin slices of biological tissues are prepared and examined under transmission optical microscopy, with and without staining. Additional image contrast is added by using fluorescent dyes, dark field optical microscopy, differential interference and phase contrast as well. Geologists works to examine mineralogical specimen of thickness less than 50 μm, observe specimen anisotropy under polarized light, optical properties and crystal phases and orientation are studied by using optical microscopy. Polymers and plastics can be viewed by reflection and transmission of

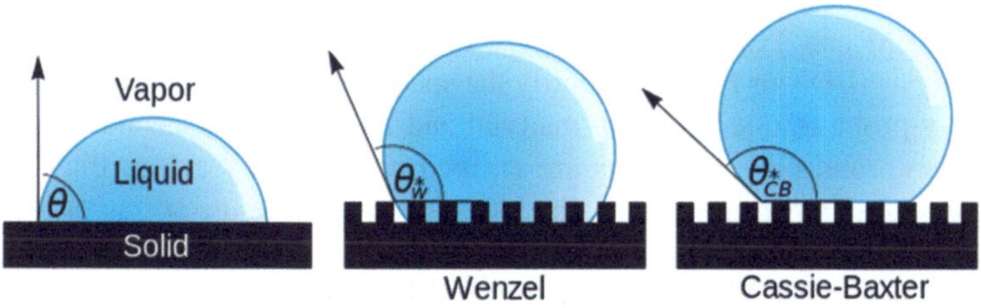

Fig. 10.2 Schematic representation of switching from hydrophilic to hydrophobic state by contact angle measurements between water droplet and solid surface. For rough surface: the droplet is in Wenzel state, where θ will change to θw. For the liquid on the top of the surfaces, droplet is in Cassie-Baxter state with an angle θ_{CB}. Surfaces with contact angle >90° are hydrophobic in nature and a hydrophilic surface has contact angle <90°. Switching between Wenzel and Cassie-Baxter states depending on surface energy and roughness was reported recently

light waves. Template based microbes and metal architectures were examined under light microscope for future electronic and sensing applications.

Recent advances show that measurement of water contact angle WCA to examine the hydrophobic effect in materials, a customized contact angle instrument, equipped with digital camera to visualize the hydrophobic lotus effect. Water contact angle (as displayed in Fig. 10.2) was recorded with customized contact angle instrument equipped with Dino lite Pro AM 413T digital camera (Tay et al. 2013). In similar manner, measurement of water contact angle on leaves is possible. Leaves have all sorts of structures and are hydrophobic, hydrophilic, super hydrophobic, oleophobic etc. This is definitely an important field with applications in advanced surfaces and agriculture.

Beside so many advance analyses byoptical microscopy, light radiations (photons) are useless to analyze the nanoworld due to limited resolution in comparison of electron wavelength, so optical microscope is not in demand for analysis at nanoscale. Improved fabrication techniques based on advanced tools and skills are required to understand nanostructures. Electron microscopes use high energy electron beam to observe matter at high resolution. Generally, two types of electron microscopes, transmission electron microscope (TEM) and scanning electron microscope (SEM) are common in use for nano analysis. Thus, we will focus on electron probe techniques in detail instead of typical optical microscope.

10.3 Electron Probe Techniques

In electron probe, analytical techniques, electron beams of shorter wavelength than typical light beam are used for imaging, chemical and structure investigations. The most familiar electron probe methods are scanning electron microscopy (SEM) and transmission electron microscopy (TEM). In SEM, the electron beam is screened across the surface. Electron probe methods are listed in Table 10.2. In most of the analysis, SEM and TEM instruments are usually coupled with energy-dispersive x-ray analysis (EDX or EDS) to identify the specimen peaks. Electron beam is used as primary probe to interacts with material and then photons, secondary electrons (low energy), and heat radiations are emitted as secondary effect. Finally, these radiations transformed into images and spectra to analyze the specimen. Electron probe methods are specifically important and useful analytical tools to analyze particles at nanoscale. Vacuum is required for all electron microscopes, scanning as well as transmission for three factors as follows:

1. A vacuum of 10^{-5} Torr is suitable to avoid the scattering of residual gases and high energy electrons within microscopic column.
2. Typically, a vacuum of 10^{-7} Torr for LaB_6 and down to $10–^{10}$ Torr for electron gun to maintain the chemical and thermal stability of electron gun within the column throughout the whole process.
3. To minimize the beam-induced sample contamination, inelastic scattering among high energy electron beam and surface absorption of residual gases as hydrocarbons forms the amorphous and carbonaceous (carbon contaminants) adhesives on sample surface.

Electrons exhibit both particle and wave like properties as confirmed by de Broglie equation that relates the momentum of an electron (a particle phenomenon) to wavelength

$$P = h / \lambda = h\nu / c$$

where P is momentum (mυ)
ν is the frequency
c is the speed of light (2.998 × 108 m·s^{-1})
h is Planck's constant (6.6262 × 10–34 J·s), λ is the wavelength. This wave-particle duality can also be expressed as λ = h/mv, where m is the mass of the particle and v is the velocity. By assuming accelerating voltage of electron beam V, electron energy can be written as eV = mv^2/2, where e is the charge on electron, it follows that $\lambda = h/ (2\ meV)^{1/2}$ or $\lambda = (1.5/ V)^{1/2}$ nm. This is an accelerating voltage approximation in electron microscope. Whereas the rest mass m_0 of electron is less than relativistic mass m, thus de-Broglie equation is corrected to
$\lambda = h/ \sqrt{2m_0eV(1 + eV/2m_0c^2)}$
AT 10 KV, many applications of SEM, the wavelength is only 0.012 nm, less than interatomic distance in solids.

10.3.1 Electron and Matter Interaction

When electron interact with matter (thick and thin specimen), number of signals are generated. These signals can be categorized into electrons and proton signals. In all cases, electrons serve as the primary high energy incident beam to work at higher

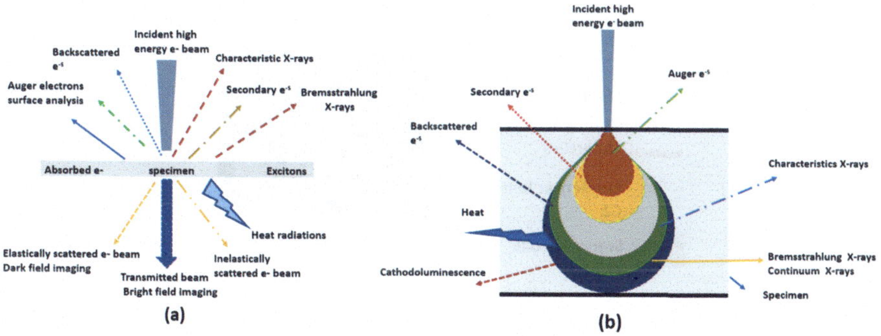

Fig. 10.3 Electron and matter interaction

resolution. When electron beam penetrates into solid material, both elastic and inelastic scattering occurs, depending on, the size of material is twice of the mean free path of an electron. For thick specimen, electron matter interaction results in a pear-shaped region is formed (Fig. 10.3a).

The interaction volume is typically within the range of 100 nm to 5 μm into the surface. This interaction volume increases with increasing electron energy and decreases with increasing specimen density (Hornyak et al. 2008, Wittke 2006). Thin specimens are analyzed by SEM images that use secondary and back scattered electrons for imaging. TEM for thin specimens, use transmitted electron beam for imaging, elastically scattered electrons for structural analysis (electron diffraction) whereas both SEM and TEM give elemental analysis by EDX simultaneously with imaging.

10.3.2 Scanning Electron Microscopy (SEM) and Analysis

Scanning electron microscope (SEM) is powerful technique to study the surface morphology and particle size and compositions of solids, crystals and highly functional materials by looking at surface or nearby surface at micro and nano scale. SEM scan the sample surface by the interaction of electron beam with surface (Goldstein et al. 1992).

Components The basic components of the scanning electron microscope are shown in Fig. 10.4. At the top, thermionic electron gun produces the monochromatic electron beam. The beam is focused to a diameter of ~50 Å at the foot of the column by a series of magnetic lenses, to de-magnify the beam, high angle electrons can be eliminated by condenser lens with objective aperture. And thus, limit the amount of current, the second condenser lens convert the electron beam into thin and coherent beam. A condenser lens is responsible for collimating the divergent beam and focusing it down the column and regulating the amount of current. Further the set of coils scan the beam within microsecond range in a TV-type raster across the specimen surface, finally the objective lens focuses the scanned beam onto desired

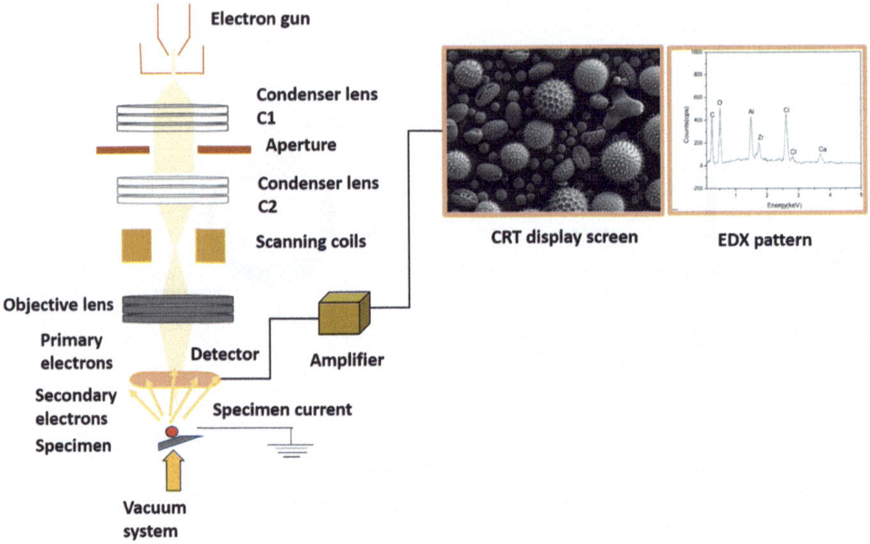

Fig. 10.4 Schematics of scanning electron microscope (SEM)

part of specimen to control and modify the beam. Upon striking the samples, interactions occur inside the sample and secondary effects emitted from the surface (e.g., secondary electrons, backscattered electrons, x-rays) and detected by various instruments including detector systems, finally, the computer programs are used to collect and process the data and display the images for analysis.

Filaments in electron gun made of two materials either tungsten (W) or lanthanum hexaboride (LaB6). Tungsten filament, hairpin style or spike like shape is commonly used due to high melting point and low vapor pressure and of low cost as well. When electron of energy Ee ~ KT, within the range of 0.2–40 keV, exit the filament, the tungsten cathode is heated by filament current. Ejection of an electron from the metal is dependent on work function of that material. Metals are found to be the best materials for electron emission as having free electrons on the surface. Energy required for electron emission is gained by heat, produced by filament current. The tungsten cathode is a wire filament bent into V-shaped tip of 100 μm radius. Figure 10.4 depicts the schematics of SEM basic components and analysis.

Image Production In electron microscopes, the electrons are emitted as a point source, deflected and focused by both electrostatic and magnetic fields. Unlike the optical lenses made of glass, electron microscopes are equipped with condenser and objective lenses controlled by a magnetic field. The electron beam exiting the anode plate of the electron gun assembly is influenced by electrostatic field. Rest of all subsequent focusing is by the electromagnetic lenses made of soft iron having zero hysteresis values. The focal length of electromagnetic lenses varies with current flows in the coils (Brandon and Kaplan 2008). An alignment of lenses within microscopic column is critically important for optimizing the better performance of electron microscope. An objective lens is probe forming lens located near the sample, thus mostly affected by misalignment. An axial asymmetry results in astigmatism of an image is caused by many factors e.g., imperfections in the magnetic lenses dielectric properties, size, position and specimen contaminants in the column. Stigmators are lenses with especial arrangement of coils to compensate the contamination and distortion in a beam.

A sample in SEM is placed below the objective lens outside the magnetic field. Samples are generally mounted on specimen stub by using a conductive adherent and can be viewed even at high temperature under both either low or high vacuum conditions. Different SEM model can examine any part of 6 inches (15 cm) semiconductor wafer, some can even tilt at an angle of 45° and also provide continuous 360° rotation. Specimen chambers to accommodate the samples 10 cm or even much larger assemblies are available in SEM (Brandon and Kaplan 2008). Electrostatic charge instability will result in unstable secondary electron emission, thus both resolution and image stability will be destroyed. Charging of non-conducting specimen must be prevented either by applying low beam voltage or by coating the sample with thin layer of conducting material (gold, gold/palladium alloy, platinum, iridium, graphite, corodium etc.). For conventional imaging, specimen surface must be electrically conductive. Non-conducting materials requires thin coating of electrically conducting material via sputtering and evaporation. Heavy metal coating increases the signal to noise ratio for the samples as secondary electron emission increases for high atomic number Z. The specimen must have to be perfectly fitted into sample chamber without any constraints to manipulate the specimen at various angles and area of interest to examine through microscopic column. And the most important is of course the vacuum system for both specimen and electron beam interaction to avoid residual gases and other contaminations that can disturb the opto-electrical system inside the column. Loose particles must be removed from samples by ultrasonic cleaning and drying before inserting into microscope require to get better images at low voltage and high resolution SEM analysis. Bulk conductivity of biological samples can be increased by OTO (O-osmium tetroxide, T-thiocarbohydrazide, O-osmium) method (Seligman et al. 1966; Malick et al. 1957).

An image of non-conducting specimen can be scanned without coating by using environmental SEM (ESEM) or low-voltage mode of SEM screening. Specimen is kept in chamber at high pressure and electron beam is pumped through the column by keeping electron gun at low vacuum. High pressure region neutralizes the charge and amplify the secondary electron signal. Charging of non-conductive samples must be prevented by adjusting the suitable conditions to keep the incoming beam, a combination of secondary electron beam and backscattered electrons at accelerating voltage 0.3–4 kV (Policarp 2005).

Biological Samples Biological samples must have to be completely dry for SEM imaging. At very high vacuum, hard samples like wood, bone, shells, dried insects can be viewed just by little modification but for soft tissues, living cells, chemical fixation is necessary to be done for stability and preservation. Glutaraldehyde and formaldehyde are used as chemical fixative (Kieman 2000; Jeffree and Read 1991; Karnovsky 1965). Fixation is usually done in incubator to avoid collapsing and shrinkage of sample caused by air-drying. Fixation is made possible by replacing water in the cell by organic solvent like ethanol and acetone etc. Finally, the replacement of these solvents is done by carbon di-oxide as critical point drying. In super critical state, carbon di-oxide is removed, thus, no liquid-gas interface present within the sample upon drying. Dry specimen is now mounted on sample stub by using a conductive double sided adhesive tape or graphite cement, and super coated with gold and palladium alloy to be examined under SEM. Some special treatment required to examine protein embedded lipid membranes (Barnes et al. 2002; Wergin and Erbe 1994).

Staining is required to increase the visibility of cellular structure of specimen by increasing the contrast with background, enhance morphological features and to preserve the specimen, especially the biological ones. Staining can be done with single staining agent and basic dyes (crystal violet). Preservation of internal and external structures and position fixing is necessary. It can be heat fixing: top reserve all morphology but not the internal structures, whereas the chemical fixing to protects cellular structure and morphology of more delicate organisms.

Image of specimen is generated by screening the electron beam across the sample surface. A reconstructed image is sent to a computer monitor cathode ray tube CRT, and viewed. During the sample scanning, an electron beam is paused at "dwell time." Number of secondary electron effects is expressed at dwell time. These secondary effects are detected, recorded, and dissipated before beam travel to next step. Magnification in SEM lies within the range of 6 order of magnitude from about 10 to 500,000 times. In an SEM, the magnification is the ratio of the dimensions of the screening on the specimen and the viewed device. Magnification is controlled by the current through x, y scanning coils, or an average supplied voltage to the deflecting plates, but not an objective lens. Magnification, brightness, and contrast all affect the image quality. Brightness of an image is controlled by the amplitude of the analyzing signals. Contrast is basically the ratio of change is signal strength ΔS between two points to the average signal strength SAvergae, can be expressed as $C = \Delta S \cdot SAverage^{-1}$.

When sample is excited by high energy incident electron beam, secondary electrons (SE), backscattered electron (BSE), characteristic x-ray photon, Auger electrons, cathodoluminescence (CL) and specimencurrent are emitted from the surface. In standard detection mode, low energy secondary electron (< 50 eV) signals (SEI) are generated from K-shell by high energy electrons incident beam from nearby surface (few nm) by inelastic scattering, produce images at high resolution. Everhart-Thornley detector is used to detect these secondary electrons. These secondary electron beam get sufficiently high energy under accelerating voltage of 400 V to 2000 V to cause light flashes emission from scintillator (cathodoluminescence) and conducted to a photomultiplier outside the column through light in specimen chamber. An amplified signal is then displayed as two-dimensional digital image can be viewed and photographed on monitor, represents

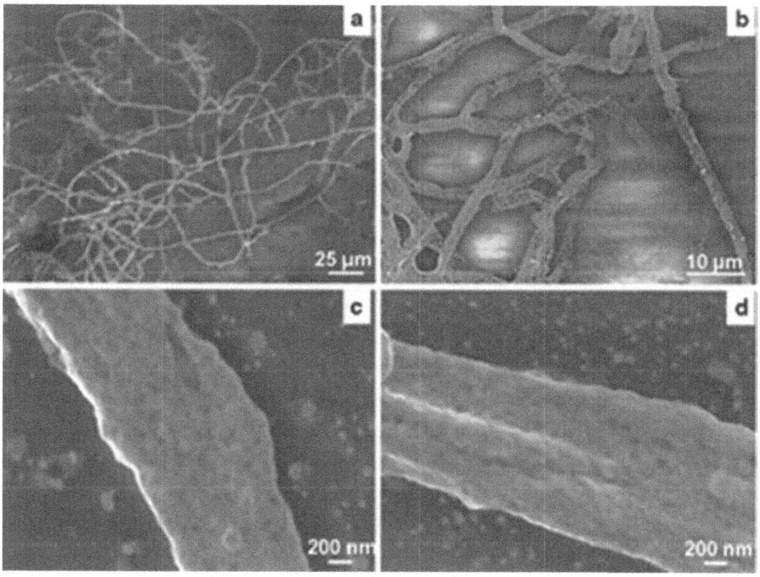

Fig. 10.5 SEM imaging: Functional microstructures of *A. niger* templated colloidal gold Microwires at different resolution with different Sizes. (**a**, **b**) mesh of wires (**c**, **d**) single microwire image (Sugunan et al. 2007)

intensity distribution in the specimen. Number of secondary electrons reaching the detector and an angle of incidence will determine the brightness of the signal. Steeper surfaces and edges produce bright images than smooth one. Whereas, back-scattered electrons of high energy travel back to surface after elastic scattering, deeper from the specimen thus produces images at low resolution than SEI. The secondary electrons generate a huge peak at low energy end of the characteristic x-ray spectrum for analytical SEM for low energy backs scattered electrons. BSE images are useful for elemental distribution as intensity of BSEis strongly dependent on atomic number Z of analyzed sample. Back scattered electrons display brighter image and used to detect the contrast in different areas of various chemical compositions (Hindmarsh et al. 2007). Especially designed "doughnut shaped" detectors concentric with an electron beam either of scintillator or semiconductor are used for BSE detection. Back scattered also produces back scattered diffraction pattern to determine the crystallography. Luminescence from minerals is investi gated by cathode luminescence. Nanoscale features and structures can be examined and analyzed by beam- injection techniques including the cathodoluminescence and electron beam induced current EBIC.

In biological samples, it is difficult to detect colloidal gold immune-labels of 5 or 10 nm diameter, in SEI (Goldstein et al. 1981). Figure 10.5 displayed the 3-D SEM imaging for fungal template based gold microwires. Fungal hyphae were used as template to deposit metal gold to fabricate highly conductive wires, a new composite of microbes and metals nanoparticles (Suzuki 2002). Characteristic x-rays are generated from inner shell core of the sample within SEM for energy and wavelength dispersive x-rays analysis. These spectra give information about composition measure and element abundance in desired sample. SEM micrographs exhibit three-dimensional images with large depth of field due to narrow electron beam, useful for understanding the surface morphology of the specimen. Magnification range in SEM images from about 10 times to more than 500,000 times, approximately, 250 times the magnification limit of the best light microscope.

Auger electron signals also exist at energies just above the secondary electron energy peak. Auger electrons gives information on surface chemistry, but it requires high vacuum. Specimen current is the basically the collection of SE, BSE and Auger electrons after impinging the incident electron beam on the sample surface. Secondary electrons have energies as high as 100 to 200 eV, but usually their energies fall below 10–50 eV. Thus, very low voltage is required for secondary electron deflection and can be collected with almost 100% efficiency. These emitted secondary electron signals are collected and examined by a scintillator-photomultiplier Everhart–Thornley detector. The secondary electrons (SEs) are first collimated by a grid with an applied voltage and then collected by detector surface. A short-life Phosphor is found to be the best to convert electron beam into photon within ultraviolet range ~400 nm. Secondary electron emission is affected by the four factors,

1. The work function of the material surface as electrons have to overcome this energy barrier in order to escape from metal to vacuum. Work function value depends on crystal structure, composition, surface adsorption and contamination films as well.
2. Energy of incident electrons and beam current. More secondary electrons are emitted at increasing incident energy, but high energy beam is scattered inelastically from the depth of sample surface, thus reduces the escape of secondary electrons yield beneath the sample surface.
3. Materials with high atomic number Z produced the large number of secondary electrons.
4. Surface topography, surface curvature and an angle of incident beam influence the secondary electron emission as well.

Backscattered electrons (BSEs) strikes the semiconductor arrays located at the bottom of the column and then detected to acquire an image. Most SEM samples are conducting, and all SEM specimens are fixed to a metal stage with conducting tape (Barmparis et al. 2015).

Extensive training is required to operate such an expensive and high-maintenance equipment. The chamber is evacuated up to 10^{-6} torr to begin the filament heating and electron emission process. The chamber is evacuated with a roughing pump. The acceleration potential is stepped up slowly. Once the beam current is detected and adjusted, the sample is introduced through column and ready for analysis. Resolution in SEM is dependent on the size of electron spot and in turn depends on the electron wavelength. An interaction volume and material volume that interact with an electron also limits the image resolution, thus spot size and interaction volume changes with atomic distances, thus reducing the image resolution in comparison with transmission electron microscopy TEM. SEM is not only used just to scan thin films or foils like samples, even the bulk material having large area. Numbers of analytical modes are equipped in the SEM to measure the properties and compositions of the variety of samples. Resolution may vary within the range less than 1 nm to 20 nm depending on the instrument model. SEM has a much greater depth of field in comparison with TEM, thus complicated surface topography can be investigated and visualize in 3D imaging (Liu et al. 2010).

Henry Mosely in 1913, first time determined the composition of material by characteristic x-rays and found that wavelength of emitted x-rays is inversely proportional to the square of the atomic number $\lambda \propto 1 / z^2$. Electron probe microanalysis EPMA was patent by James Hillier in 1943. EPMA is designed as non-destructive x-ray technique and capable for ultrafine spot focusing, down to 1 μm, optical imaging and sample positioning as well. The EPMA has an excellent spatial resolution capability of 1 μm within detection limit of 100 ppm. EDX detectors are included for fast analysis. X-ray energy in the range of 0.1–15 keV is typically analyzed. Images are produced by collection of backscattered electrons. X-ray wavelength range for wavelength dispersive spectroscopy, WDS is from 9.3 to 0.087 nm. Large specimen size can be accommodated with dimensions of 100 × 100 × 50 mm (height) and an analyzable area of 90 mm^2. The probing current ranges from 1 pA to 10 μA. The backscattered electron image detection limit is 6 nm with a working distance (WD) of 11 mm and under 30 kV accelerating voltage conditions (Donovan 1992).

Further, to avoid attenuation of electron signal by air-gas atmosphere, high vacuum is required for SEM imaging. For example, the samples like wet biological samples, oil-bearing materials, liquid transport, melting of alloys and chemical reactions that produces significant number of vapors must be cryogenically frozen before analyzing. In late 1980s, environmental SEM with especially designed secondary electron detector was developed is capable to operate at low-pressure environment, 1–50 Torr and high relative humidity, up to 100% (Mancuso et al. 1998; Danilatos 1990).

ESEM is very useful to analyze the non-metallic and biological stuff. No carbon, gold and gold alloy etc. coating is required for plastics, elastomers and biological entities. X-ray analysis also become difficult with coated materials, may also destruct some surface features, instead of that ESEM make x-ray microanalysis possible for uncoated and non-conducting samples. SEM can also analyze the samples in transmission mode by simply using a detector below a thin specimen (Klein et al. 2012). SEM basically produce grey-scale images, as single value per pixel; corresponds to total number of secondary electrons reaching to the detector during small period of scanning time, when the beam is targeted to the (x,y) pixel position (Burgess 1987). Different techniques are used to produce the colored images artificially, as false coloring to distinguish the transition phases, usually on BSE images, photometric 3D rendering from a single SEM image to render the surface texture, aesthetic effect for real appearance of the sample and also manually with photo editing software for feature detection or object-oriented segmentation (Brożek-Mucha 2014; Nawa et al. 2014; Hortolà 2010). In addition to 2D imaging of specimen, SEM analysis provide qualitative and quantitative information about particle shape, size, surface morphology, texture and topography as well as chemical composition of the specimen by collecting secondary (2° electron) and backscattered radiations. In recent years, the electron beam nano-lithography, and focused ion beam techniques are used to perform in-situ advanced and précised nano-engineering. Within large scale development in SEM analytical technique, still some limitations, for example, conductive coating of samples, even a thin layer but may cause specimen aircraft. Traditional SEM also requires high vacuum test environment, limited resolution even at high pressure. Electron beam scanning cause the damaging of tested sample is another drawback, need to be considered. Hence with all these limitations, still there is plenty of room in for researchers and scientists.

10.3.3 Transmission Electron Microscopy and Analysis (TEM)

In Transmission electron microscopy, electrons are transmitted through a thin specimen, formed image after interaction of electron beam and specimen is focused and viewed on the photodetector like photographic film, fluorescent screen or a sensor such as charge-coupled device CCD. Specimen must be thin so that electron can penetrate through material for analysis. TEM operate on the same principle as SEM, the filament (made of tungsten and lanthanum boride as in SEM) in thermionic electron gun is heated under an applied potential to produce images. The main difference in TEM and SEM is that the detector is a phosphorous plate and an accelerating voltage in TEM is far greater: 300 kV compared to ~50 kV. The wavelength of the electron beam is in the picometer range, 6.13–2.24 pm. TEMs have a projector lens system at the base of the column, thus projected image is observed on phosphorous screen.

Components A schematic version of a generic TEM is shown in Fig. 10.6. A field emission gun is used to accelerate electrons between 100 and 400 kV. The electron beam is accelerated by the anode plate and then collimated through an aperture. The accelerated electron beam is transmitted through the sample, typically are 1000 atoms or less in thickness (e.g., >100 nm) TEM consist of three series of lenses, the condenser lenses to form the primary beam of electrons, the objective lenses and projector lenses to focus and expand the beam on detector, a phosphorous screen or imaging film etc. Transmitted beam is focused by an objective lens, channeled through another aperture, and then projected onto the fluorescent screen. The CCD camera or photographic plate captures the "shadow image" of the specimen depending on its component density. CCD cameras are integrated circuits, composed of close-packed photodiodes that convert the emitted photons into fluorescent images. These photon signals are produced by electron impinging onto the specimen. In order to record and display an image, the phosphor screen is tilted for larger and open view to the CCD camera. CCD camera follows the photoelectric effect is invented by Willard Boyle and George Smith of Bell Labs in 1969. The first CCD camera was used in a TEM in 1982 (the 100 × 100 pixel device directly exposed to

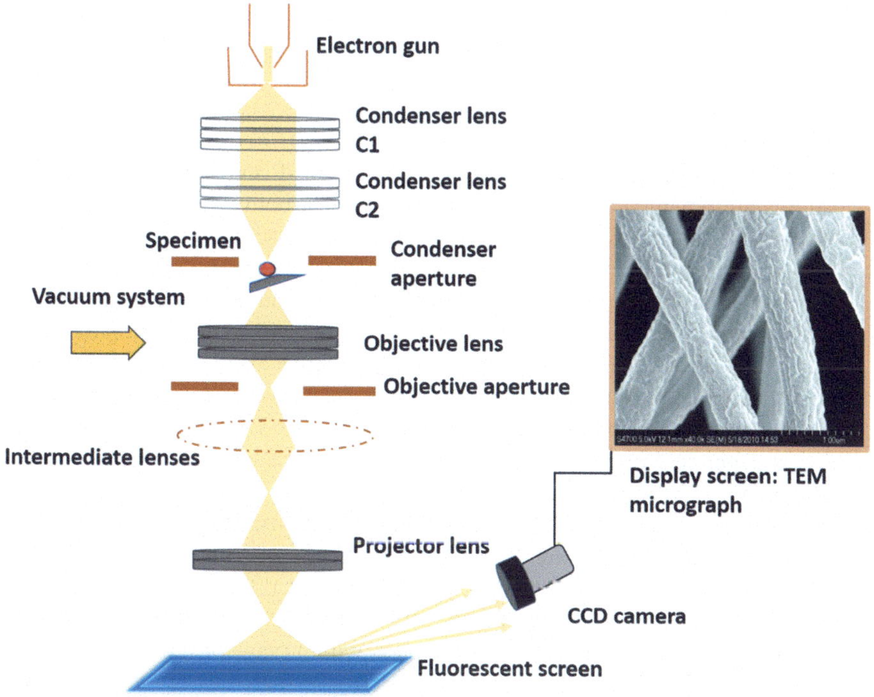

Fig. 10.6 Schematic representation of transmission electron microscope (TEM)

100-keV electrons). Some issues were observed by direct detection, thus J. C. H. Spence and J. M. Zhou proposed an indirect detection method by suing scintillation screen with an optical coupler (Ellisman 2005; Spence and Zhou 1998; Boyle and Smith 1970). Now CCD technology is modified with 1000 × 1000 pixel arrays for TEM imaging. The resolution of a good TEM is ca. 0.2 nm, on par with the distance between two atoms and the atomic radii of some heavy metals. In 2005, Carl Zeiss SMT introduced the resolution of 0.07 nm with an experimental ultrahigh resolution 200-kV field emission gun transmission electron microscope (FEG-UHRTEM (ultra-high resolution)) that was near to theoretical limit of TEM analysis.

Image Production TEM produces the images with better resolution and fine detail up to the level of single column of an atom by following the de Broglie wavelength. Generally, four types of images are produced by TEM:

1. Transmitted electrons are responsible for generating a bright field image of the specimen, a most common operation mode for TEM analysis. When the specimen is placed just above the objective aperture, only the transmitted electron beam is allowed to pass down the column by excluding all diffracted and inelastically scattered electrons and form an image on the fluorescent screen.

2. A low-resolution dark field image is obtained by adjusting the objective aperture to one side or another of the main axis of the column and an inserted metal plate is used to block the major beam. Beside the drawback of spherical aberration in this arrangement, a quick investigation of specimen is observed by formed diffraction pattern.
3. A high-resolution dark filed image can be formed and electron diffraction patterns are obtained to identify the lattice defects, both type and positions by carefully adjusting the sample orientation (Hull and Bacon 2001).
4. High resolution lattice images are formed in a high-resolution (HR)- TEM mode of operation also called as phase contrast. Both real and reciprocal space can be observed in this imaging. When an aperture is opened to allow to pass many beams, images are formed by interference of the direct and the diffracted beams. TEM images are used to observe the crystal orientations, chemical identity, electronic structure and phase shifting (Kirkland 1998). Direct absorption based images depending on thickness and composition of sample are observed at low magnification. At higher magnification, image intensity is based on the complex wave interaction and requires keen and expert analysis. Magnification of TEM can be changed by modifying the current through the coil, quadrupole and hexapole lenses. The specimen holder is inserted into vacuum to minimize the surrounding pressure by using airlocks on specimen stage. Grid of specimen holder is ~ 3.05 mm in diameter and mesh thickness within range of few 100 μm, upon which the sample is placed. Commonly used grid materials are copper, gold, platinum or molybdenum. The sample is placed on mesh within the diameter of ~ 2.5 mm. To accommodate the sample for region of interest, TEM stage provides translation symmetry on XY plane and has at least one rotation degree of freedom for the sample to adjust for Z height within the holder. Modern TEM are especially designed with double-tilt sample holder to provide two orthogonal rotation angles. To promote specimen insertion into holder without damaging, the side-entry and top entry version is introduced for specimen stages in TEM.

The specimen with thickness of few 100 nm or less is usually placed on a copper grid that is less than 100 μm thick and 3 mm in diameter unlike neutron or x-ray diffraction methods. Thickness of samples must be comparable to mean free path of electrons, as electron beam has to be transmitted through the sample of less than nm thickness. Variety of techniques are introduced observe the sample near to natural state with all feature preserved to avoid incomplete interpretation of data, number of preparation methods were introduced e.g. tissue sectioning, sample staining, dehydration, infiltration, mechanical milling, chemical etching etc. These all preparation methods are material specific under TEM analysis. Materials like powders or nanotubes are already thin and electron transparent, so can be prepared quickly just by direct deposition in acetone or other volatile solvents (uranyl acetate) onto carbon-coated copper grids. Most of the biological samples are structurally weak, hydrated and semi-transparent to electron. Initially it was considered that biological samples cannot be examined under TEM due to high vacuum, generated heat and of course the penetration depth of electrons.

Biological Samples TEM analysis of biological samples requires the staining with heavy metal dyes that contain osmium, uranium, gold or lead. Staining of biological tissues with high atomic number enhances the image contrast. In order to observe the sample near to natural state with all feature preserved to avoid incomplete interpretation of data, number of preparation methods were introduced, for example, dehydration, infiltration, embedded and sectioning Schematics of specimen preparation may be observed in Fig. 10.6.

A process of fixation by adjusting the influencing factors like change in pH, temperature, osmotic pressure etc. (Amzallag et al. 2006). For hard inorganic samples, sawing, grinding, and polishing are done to make it to several microns to required level. The polished sample is then placed in an ion mill to undergo more finer level of thinning by bombarding the argon ions and remove the material according to the accelerating voltage of the ion mill system. Other than mechanical polishing, chemical and electrochemical thinning and reactive ion etching techniques are also used to reduce the thickness of samples for TEM analysis. Tissue sectioning is a technique used to obtain thin, minimally deformed sections by glass or diamond edge. Tissue samples under analysis need to be coated by several nm thin conducting layer of carbon (Baram and Kaplan 2008). Embedding of sample in a polymeric resin and sectioning with a microtome by the action of a diamond knife is another method to produce thin films with nanometer thickness. Floated specimens on the water are collected on a copper grid, dried, and placed into the TEM sample holder in the middle of the column. The thickness of sectioned floating samples is estimated by observing the interference colors. Films that are thinner than a few hundred nanometers appear gray as they float on the water collector. Sections with thickness range of 200 nm to 1 μm display all the colors in the spectrum (Phillips 1961).

Of course, the limitations are still there for TEM analysis as for small size of samples, the field of view is relatively small that may cover the samples section for analysis, instead of whole. Extensive sample preparation techniques further the structure and morphology of the samples changes drastically under exposure to electrons with extremely high energy and damage the samples, particularly the biological ones. TEM analysis requires high-vacuum, typically of the order of 10^{-4} to 10^{-7} Pa or higher to increase the mean free path that is costly to operate and to maintain, providing a big hindrance to conduct nanoscience research. This vacuum can be achieved by adjusting the voltage difference between cathode and the ground state. Multiple pumping systems and air locks are attached with TEM to achieve such high value. Initially, a rotary pump provides a low or rough vacuum, then a turbo-molecular or diffusion pump establish high vacuum level necessary to start an operation.

Beside all that limitations, the advantages of the TEM technique are numerous. All type of samples, electrically insulating, semiconducting, or conducting is able to be imaged by TEM. Incredible resolution capability of TEM allows atomic level inspection very well. The main use of TEM images on biological scales is to examine the structure, properties and composition of the specimen like cells, tissues, cell fractions and isolated macromolecules at submicron levels, as cannot be done by other techniques and ways. For heavy atoms, sites appear darker due to intense

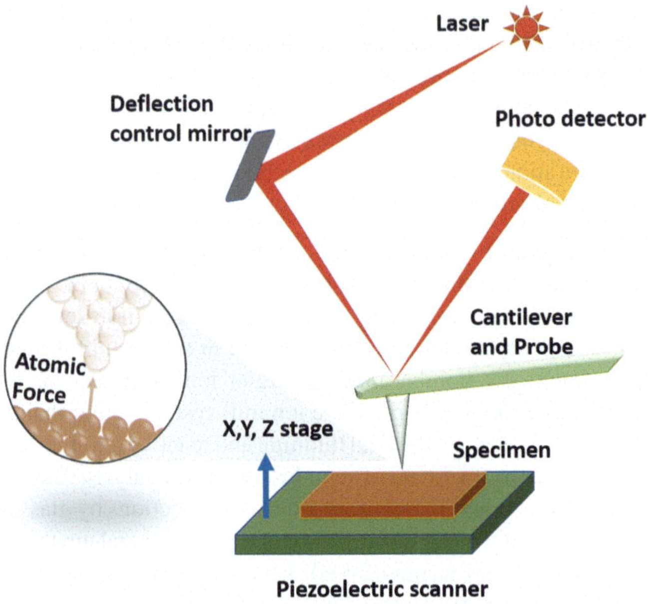

Fig. 10.7 Schematic representation of atomic force microscopy (AFM)

scattering and absorption of electrons. Particle sizes measurements, direct observation of crystal structure and orientation is possible from acquired pattern. Kakuchi bands consists of light and dark lines are formed by inelastic scattering of electrons are used to study the atomic spacing of the samples. TEMs analysis is important for medical and cancer research as well, microbiology, nanotechnology, material science and semiconductor applications.

3-Dimensional imaging (Fig. 10.7) is visulized to reconstruct the ornamentations and wall structure of pollen grains by using focused ion beam SEM (FIB-SEM) analytical technique. Study of spores and pollens are imprtant for understanding the botanical affinities and phylogeny for establishing ancestral pollen states in early flowering plants. For studying ultrathin sections of spores and pollen grains, ultra-micrometery is performed by using TEM (Villanueva-Amadoz et al. 2012).

10.4 Scanning Probe Techniques

Three-dimensional imaging of specimen surfaces can be visualized by scanning probe microscopes at atomic level. Two basic types of scanning probe microscopes are commonly used in nanoscience. The first is scanning tunneling microscope (STM), developed in 1981 by Binnig and Rohrer (Binnig et al. 1982). In the STM technique, analysis being done by monitoring the tunneling current between the probe tip and atoms or molecules of surface substrate. Thus, STM technique is suitable to analyze electrically conducting samples. Second is atomic force microscopy (AFM) is introduced in 1985 to analyze the insulating and electrically conducting samples. In AFM, the size of the force (~ 1μN or less) between a probe tip and the atoms of substrate surface is monitored. Both STM and AFM are equipped with a probe tip fastened to a cantilever, a scanning (motion) mechanism, and a detector system. In order to avoid scanning probe from external vibration, nitrogen-gas regulated suspension and motion-damping eddy currents are induced by using magnetic plates. Computerized feedback system controls the piezoelectric scannersat nanometer level to move the probe across the specimen surface. In case of AFM, van der Waals attractions exert force on the probe tip and tunneling flows in STM. Sharper the probe, better the resolution of surface features. Carbon nanotube probes are found to be very promising for better resolution in past few years.

Scanning Probe Tips. The probe is usually made of tungsten for STM and silicon or Si3N4 for AFM. Atomic probes must be sharped and positioned as close to the sample surface as possible, to determine the material features. In early times, profilometers equipped with diamond tip attached to coil in an electric field. Screening the profilometer tip across the surface induce current that relates to surface features [43]. Modern profilometers for example, Ambios XP-2 can measure the surface roughness sample surface. The AFM is powerful and versatile equipment for surface analysis in the field of nanoscience, a modern version of prosaic profilometer of the 1920s.

Piezoelectric Materials: Piezoelectricity is production of electricity in certain materials by applying mechanical stress. Pierre Curie and his brother, Jacques, discovered that whenever, pressure is applied to crystal orientation, an electricity is produced, the phenomena is called as piezoelectricity and vice versa (Scalf and West 2007). Quartz, ceramics, and Rochelle salt are commonly used piezoelectric materials. Piezoelectric devices include dielectric materials that actually convert mechanical stress into electrical signals. Devices such as cartridges, microphones, radios, and strain gauges make use of piezoelectric materials. For scanning of specimen under AFM and STM are equipped with electrochemical piezoelectric ceramic transducers to provide mechanical motion. A standard piezoelectric crystal can expand about 1 nm per AC volt applied, but vary for different types of materials. Piezoelectric transducers consist of hundreds of layers to manage all variable motions required for surface imaging. For example, motion of 1000 nm per volt for 1000 layers of transducer, whereas motion of 0.05 nm for 50 V is possible (Bamparis et al. 2015). Two type of scanning modes are possible, either sample is moved relative to the probe tip or the probe is rastered across the surface by keeping the sample stationary. In one configuration, the piezoelectric device controls the x and y displacement of the sample and the z motion of the probe tip.

10.4.1 Atomic Force Microscopy AFM

AFM is a powerful and very simple mechanical tool for imagining and analyzing the surface features at atomic scale in the field of nanoscience, under normal conditions without requiring any especial environment and high energy beams. AFM is versatile and suitable technique to analyze the biological samples due to weak forces between the probe tip and surface. Atomic force microscope is equipped with a cantilever, sharp probe tip of 50 to 20 nm or less. The mechanical deflection of cantilever within descending tip gives information about surface structure. Distance between tip and sample is kept fix in order to keep the force constant between them by well-defined feedback mechanism. Laser light is focused on top surface of the cantilever into the photodiode detector. The incoming signal from the detector is directly proportional to the cantileverdistance (Fig. 10.7).

Split photodiode in AFM measure the differences in the reflected beam and thus recorded the changes in surface topography and produce 3-dimensional grid-like image of surface. Basically, the AFM analytical technique keep balance between attractive van der Waals and repulsive electrostatic forces between the probe tip and the surface. Theoretically, the magnification is of the order of 10^9, with atomic scale resolution. AFM can also be used to image the objects as large as tens of microns or more size. For example, the tobacco mosaic virus is 300 nm in length and the human hair is 50–100 μm in width, both can be imaged by AFM. Different cantilever and tip configuration are available for AFM to operate in three different mode of operation: contact, taping and noncontact as listed in Table 10.3. Mostly the single crystal silicon is used to construct the cantilever. The length of AFM contact and tapping mode cantilevers is between 100 and 500 μm; the width is between 25 and 40 μm, and the thickness is between 1 and 10 μm. The resonant frequency of these cantilevers lies in between 10 and 300 kHz and the spring constant between 0.1 and 50 $N{\cdot}m^{-1}$. An average diameter of the probe tip ranges from a few nanometers to ~50 nm. A force between the cantilever and electronic interaction of surface sample is measured by a sensor in AFM. The force simply described by the Hooke's law $F = -Kx$, where F is the force, k is the "spring constant" of the cantilever, and x is the distance of the deflection of the cantilever.

Resolution in AFM varies within radius of curvature and tip size. Image resolution in AFM can be visualized in three dimensions: the x-y plane, in plane, typical of optical microscopes and in the perpendicular (z) direction. At contact mode of AFM, radius of curvature is quite large of the order of 50 nm and side of probe makes contact to surface before a tip. Resolution increases as close the tip to surface. The lateral resolution of a AFM is ~1.5 nm and vertical resolution of a is even better ~0.05 nm. In recent years, both multi walled and single-walled carbon nanotubes are found to be promising as probe material. Use of carbon nanotubes as tip material displayed excellent mechanical and vibrational properties. Specimen is fixed on the flat surface of metal plug in AFM with adhesive spot. Position the probe is adjusted over the sample surface by an optical microscope. The image of the cantilever as it in close contact with sample is visible on a cathode ray tube CRT screen. An

Table 10.3 AFM operative modes, configuration and analysis

AFM operative mode	Configuration	Analysis
Contact mode	The probe-cantilever assembly applies a constant force to the surface; the force constant (or spring constant) is <1 $N{\cdot}m^{-1}$. The deflection of the cantilever is due to topographical changes characteristic of the surface. The mechanical deflection is translated into an optical signal by reflection of aligned laser light from the top surface of the cantilever into an aligned calibrated dual photodiode collector. Due to intimate contact with the surface, large lateral forces can influence the probe. Surface contamination, electrostatic forces, and heterogeneous surfaces are able to impact the action of the probe. AFM is capable of imaging nanomaterials in air, vacuum, and liquids.	Samples with hard surfaces are appropriate for contact mode AFM analysis. AFM is capable of imaging insulator, semiconductor, and conductor surfaces. Qualitative information includes three dimensional visualization and material sensing by phase contrast (chemical contrast of features with the same z-displacement). Quantitative information includes topographic mapping, particle and pore size, particle and pore morphology, surface roughness and texture, particle count, size distribution, surface area distribution, and volume–mass distribution (Scalf).
Tapping mode	Strong repulsive regime oscillating probe (>100 kHz) Intermittent contact with surface lateral forces reduced significantly amplitude and phase imaging	Three-dimensional topography used for soft samples or those weekly bound to surface biological materials DNA carbon nanotubes
Non-contact mode	Weak-attractive regime oscillating probe can be applied with water layer	Soft tissues and surfaces

automated feedback system and computerized system is installed with AFM to manage the probe assembly lower to the sample surface and to guide the tip down the z-axis. 3-dimensional image is the compiled results of line scans. Scan area, probe surface and scanning rate are control parameter for images.

Bio-AFM is becoming an advanced tool for surface imaging and topography, for example multi AFM images are shown in Fig. 10.7. It has wide range of applications in soft matter and life sciences research.

10.4.2 Scanning Tunneling Microscopy STM

STM scanning tunneling microscope is an imaging technique based on strength of tunneling current exist between probe tip and substrate surface. Specimen surface must be conducting and closer to probe tip but avoiding the physical contact. Electron tunneling is the function of probe tip, applied voltage and density of states

(DOS) of specimen. Dual nature of an electron, permit them to tunnel beyond the surface of the material. Such quantum tunneling occurs; when an electron of low kinetic energy exists on other side of high energy potential barrier against the fundamental laws of classical physics. Tunneling is the penetration of an electron into a classically forbidden region. In tunneling, the finite probability of position of electron is described by wave function. The principle of quantum tunneling was also used to explain exponential radioactive half-lives decay rates. Small changes in the distance between the probe tip and the substrate surface may cause the large changes to occur in the tunneling current. STM produces three-dimensional images with atomic scale resolution. Magnitude of current is quite different or SEM and STM for sample analysis.

Direction of current flow is determined by the applied potential. If the tip is at negative voltage with respect to the substrate, then electrons will flow from the substrate to the probe for positive tip biasing, the electrons flow from the probe to the surface.

STM main components are metal tip, piezoelectric scanner, current amplifier, applied biasing and feedback current mechanism as depicted in Fig. 10.8. Magnitude of tunneling current from tip to the sample is dependent on applied biasing, exponentially on the distance between probe and surface. Feedback is normally based on the constant current and measure the height on the surface.

Scanning Tunneling Spectroscopy STS is complementary technique to a complementary method to STM. STS spectrum actually generates a map of the localized electronic structure of a specimen with atomic level resolution. Basically, the change in tunneling current is measured as a function of voltage in an I–V curve of a selected section of the sample. Two operation modes are commonly used, in one; current is held constant during a scan while height varies and recorded. In another, current varies by keeping sample height fixed. Work function and conductivity of material plotted and measured against applied V by the STS technique.

10.5 Spectroscopic Techniques

Spectroscopy is a branch of science concerned with the investigation and measurement of spectra produced by the interaction of matter and electromagnetic (EM) radiation." A spectrum, displays the intensity of radiation, emitted, absorbed and scattered by the specimen against photon wavelength, energy and frequency. Spectroscopy can be used to investigate broader spectrum for analysis involving no electromagnetic radiations to some extent. Matter and radiation interaction may produce different phenomena like absorption, emission, fluorescence luminescence, reflection, diffraction and scattering, may be elastic with no energy loss and inelastic at different energy values. For single atoms, these choices are limited to electronic transitions. Total energy during molecular absorption can be interpreted as follows

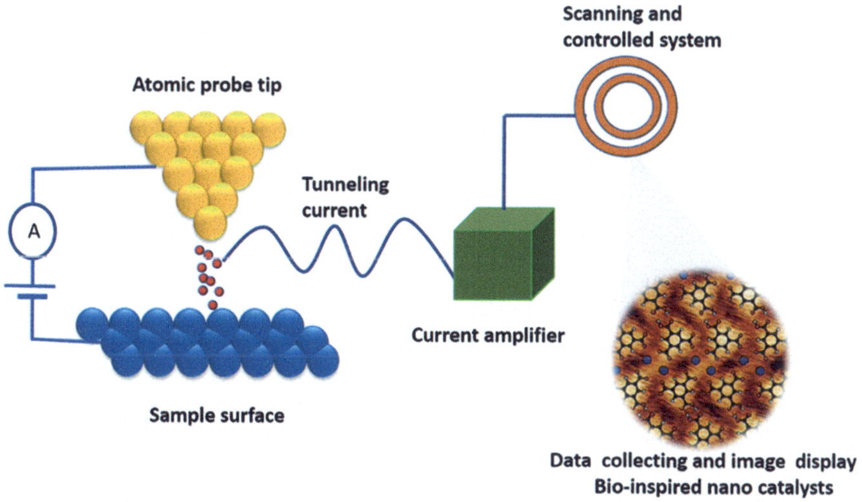

Fig. 10.8 Schematic representation of scanning tunneling microscope (STM)

$$E\,\text{total} = \Sigma\,\text{interaction} = E\,\text{electron} + E\,\text{rotation} \\ + E\,\text{vibration}\ (\text{non}\quad \text{radiative interaction mechanisms})$$

Sigma bond ($\sigma \rightarrow \sigma^*$), pi-bond ($\pi \rightarrow \pi^*$) and electron pairs exhibit atomic and molecular ground to excited state electronic transitions. Nanomaterials exhibited quantized energy transitions between discrete quantum and continuous bulk domain at atomic and molecular level, especially in small clusters. For example, larger colloids are able to scatter light but quantum dot clusters have highest occupied (HOCOs) and lowest unoccupied (LLUCOs) cluster orbitals [2].

The concept of resolution in atomic force microscopy (AFM) and other scanning probe methods is quite different from that of radiation-based microscopy. First of all, AFM imaging is a three-dimensional technique. Resolution limits of STM or AFM samples are quite impressive with resolution on the order of the atomic scale. The sharpness of the tip, the geometry of the sample surface, and the type of scanning mode all play a role in determining the resolution value.

10.5.1 UV-Visible Absorption and Emission Spectroscopy

Ultraviolet-Visible absorption is valuable technique for identifying and characterizing the nano materials having optical properties, sensitive to size, shape, concentration and refractive index near the particle surface. UV-Visible spectrum lies within the range from 300 to 800 nm. Metal nano particles display various effects/ phenomena, when interact with UV-visible light like plasmon oscillations on the surface, exciton pairs in semiconductors, size-dependent fluorescence (semiconductor quantum dots, single-walled carbon nanotubes), and interference colors (porous alumina or titaniathin films), electron transferring in supramolecules like chlorophyll and many more. It is well known that light can be transmitted, absorbed or reflected. Transmittance T is the ratio of radiant power I emerging from a material to the incident radiant power Io.

$$T = I / Io$$

Size, shape, composition and aggregated state of sample decide the relative percentage of scattering and absorption of light by specimen. As a general rule, smaller particles have higher percentage of their extinction co-efficient due to absorption. In UV-visible spectroscopy, specimen absorbs or reflects the light in the visible region of the electromagnetic spectrum. Absorption actually measure the transitions from ground to excited states of particular material. A beam of ultraviolet UV and visible light separate out into discreet colors of characteristic wavelength by a prism or diffraction grating. A monochromatic beam then split up into two beams by semi-mirrored plate. One beam passes through the cuvette having the solution mixture (sample must be dissolved in non-absorbing solvent) and the other one passes through the identical cuvette having the pure solvent only, set as references. Intensities of solutions are measured by the digital detectors. Intensity of reference beam I_0 bears almost no absorption, whereas the specimen beam intensity I shows drastic changes by light absorption and transmission. Finally, the spectrometer scans all wavelengths components over short period of time in described pattern. Scanned region lies within the range of 200–400 nm for ultraviolet (UV), and 400–800 nm for visible light. Excitation of non-bonding electrons from lower energy states to higher energy states anti-bonding molecular orbitals occurs by the absorption of UV and visible light. The absorbance is directly proportional to the concentration of the absorbing species in the sample solution and the path length,

$$A = -\log 10T = \; bc. \; (\text{Beer} \quad \text{Lambert Law})$$

where ε is the molar absorptivity ($L \cdot mol^{-1} \cdot cm^{-1}$), b is the path length of the sample in centimeters, and c is the specimen concentration. UV-visible spectrophotometer is used for both qualitative and quantitative analysis. Absorber concentration can be measured by keeping the path length fixed. Extinction co-efficient and the calibration curves are used to check that how fast the absorbance changes with concentration. In measured spectra, bonding type and functional groups of specimens can be identified by corresponding wavelengths. Solution concentration, temperature, pH, solvent nature and interfering substances can affect the observed spectrum. Width of spectrophotometer slits will also change the spectrum.

Monochromatic light within wavelength range of 190–800 nm passes through the sample and transmitted light is detected to analyze the sample at different energies by varying the wavelength. Conversely the amorphous silica does not absorb light in the visible range of spectrum, instead silica colloids scatter the light, exhibited Rayleigh scattering. Scattering is very sensitive to aggregation state of sample, for example, optical properties of silver nanoparticles become different when particles aggregate, conduction electrons become delocalized to make bonds with neighboring atoms. As a result, surface plasmon resonance at low energies shift the scattering and absorption peaks to red end of spectrum at longer wavelength. UV-Visible spectroscopy is always found to be an easy and reliable technique for monitoring the stability of nanoparticle solutions. Intensity of extinction peak decreases upon destabilization of particle, and broaden at longer wavelength due to particle aggregates.

UV-visible spectroscopy is always being widely used technique for structural characterization of silver nanoparticles synthesized by plant extracts and also the absorption of plant pigments etc. The optical spectrum of gold solution obtained with four plant extracts. Abbreviations: *Peg, Pelargonium graveolens; Pug, Punica granatum; lc, Lippia citriodora; sO, Salvia officinalis*; aU, arbitrary units and an absorption of plant pigments are investigated respectively (Fig. 10.9) (Zach et al. 2014; Mehta 2012). Varied colors of gold, other metals and colloids are due to higher order Plasmon excitations, Mie scattering and dipolar Plasmon excitations. Dipolarplasmon resonance occur, if the radius of nano metal is approximately 1/100 of the wavelength ($r < 0.01\ \lambda$). According to Drude optical model, the collisions of electrons of smaller particles with surface boundary become prominent and thus produce sound surface effect. This process increases the energy of the electrons and creates a blue shift in λ max. Whereas, for the larger particles, scattering and higher order plasmon resonances are excited and thus λ max shifts towards red end. The size-dependent emission of visible light from semiconductor quantum dots is fascinating research for the past few years.

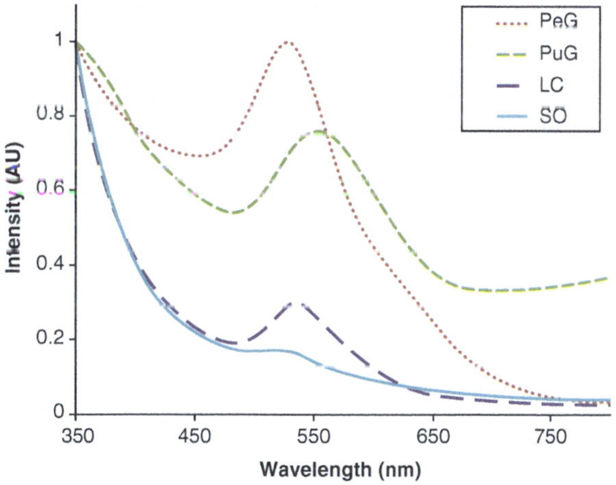

Fig 10.9 Normalized ultraviolet-visible spectra of the gold solutions obtained with the four plant extracts [48]

10.5.2 Infrared and Raman Spectroscopy

Raman spectroscopy is used to investigate vibrational, rotational and low frequency modes in materials. Raman spectroscopy, near-infrared, infrared, far-infrared and Surface enhanced Raman infrared (SERS) are relevant characterization methods. IR and Raman techniques are based on asymmetrical and symmetrical vibrations respectively, basically the two complementary methods. Scattered radiations can be categorized into three types. First is Rayleigh scattering have similar wavelength to incident beam and more intense than other two types. Rayleigh scattering (elastic) involves no change in the polarizability of the molecule. However, a small portion of the incident light exhibited inelastic at the frequencies that are different from incident photon beam, the result is exhibited in the form of vibrational, rotational, or electronic energy changes in the molecule.

Other two are Stokes and Anti-Stokes, In Stokes lines, Raman scattered beam is at longer wavelength (lower energy) than excitation wavelength, whereas, in Anti-stokes, the wavelength is shifted to shorter wavelength and higher energies. Magnitudes of Raman shifts are independent of the excitation wavelengths (Fig. 10.10). Scattering, nanoscale phenomena only occurs when particle sizes are smaller or comparable to incident radiations. If particle size is larger than incident wavelength, then reflection and refraction occur. Raman spectroscopy is the inelastic scattering of monochromatic light, usually the Laser light is used as excitation source to avoid the absorption, photo-decomposition and fluorescence. Raman effect links the vibrational frequencies to the energy differences between incident and scattered light. However, Raman analysis only displays the polarization modes transitions of molecules. Generally, all the polarizable molecules in plant cells

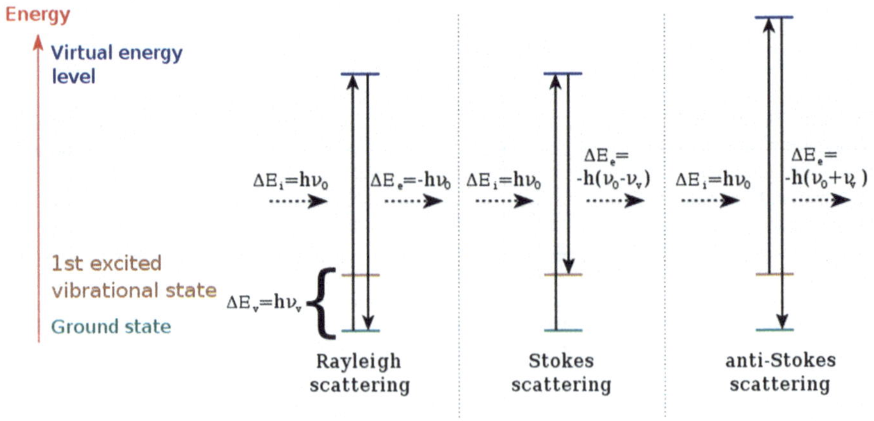

Fig. 10.10 Light scattering modes: In Rayleigh scattering, no energy exchange was observed, incident and scatted photon have same energy. In Raman scattering stokes, energy of scattered photon is lesser than incident photon, whereas, in anti-stokes, energy of scattered photon is greater than energy of an incident photon

generates the Raman spectra (Blackie et al. 2007). Raman spectrum gives very specific information about chemical bonding in molecules, microanalysis and imaging. The fingerprint region of molecule lies within the range of 500–2000 cm^{-1}. Basically, the Raman effect is based on excitation of molecule from ground state to virtual energy level, and relax into a vibrational energy level, by generating the stokes. Whereas, if the molecule begins with vibrational state to ground state after scattering, have shorter wavelength and high energy, generates anti-stokes. Raman scattering intensity is dependent on change in polarizability and Raman shift is equal to involved vibrational level.

Selection rule for Infrared IR absorption is change in dipole moment, good for polar molecule as exhibit very strong absorption in water molecules. IR signal intensity is good and have spectral range within 400–4000 cm^{-1}. Raman spectroscopy is change in polarizability, good for non-polar molecules, thus shows week absorption in water. Spectral range lies within 200–2000 cm^{-1} with week intensity signal in comparison to IR. Raman scattering is always mixed with fluorescent background thus signal processing is difficult to handle. Raman is used to detect in-organic system, especially the functional groups. Raman spectra of scattered radiation are usually measured at an angle of 90° by visible spectrometer.

The surface enhanced Raman spectroscopy (SERS) is surface sensitive technique that depends on rough metals and nanostructures that are able to enhance signal intensity by multiple orders of magnitude (10^{10} to 10^{11}), detection sensitivity can be enhanced upto 6–10 order of magnitudes over Raman spectroscopy. Two types of photon scattering, elastic scattering having no energy loss, is called Rayleigh (Rayleigh–Debye) or Mie scattering. Plant components can be detected even in low concentration by providing SERS substrates. SERS basically on the surface plasmon resonance of the nanometal particulates. SERS is widely used to detect pesticides and plant chemicals coupled with computer based analysis to identify the products. Single molecule detection may also possible by SERS (Schulz 2014).

Various Raman and infrared spectroscopic techniques play a prominent role for qualitative and quantitative analysis in plant cells for last 20 years as well in horticulture and agriculture fields. Vibrational measurements can be performed directly on the plant cells or on the plants fractions, separated from material by solvent extraction or hydro distillation. Raman spectra are unique for each material, can be said as finger print. Raman analysis doesn't require any vacuum for analysis, thus save the expensive vacuum equipment. It is non-destructive and fast analysis technique. Explosive drugs, other hazardous chemicals, origin of paper, pigments and inks materials may also be detected by Raman analytical technique by book of Kells. plant metabolic fingerprinting by Raman analysis provide information about the change in response to the genetic background, various plant diseases or influences by the environment (Schulz 2014). Usually, both IR and Raman band spectrum provide information about chemical compositions, discrimination of species, even chemotypes among same species for investigated specimen. Raman vibrational modes are also useful for protein and amino acids, fatty acids and lipids, few reports on carbohydrates, primary and secondary metabolites, detection in plant tissues. Now a days, portable Raman spectrometers are available that require only few milligrams or microliters of samples for investigations. Surface enhanced Raman spectroscopy (SERS) to detect pesticides and plant chemicals coupled with computer based principal component analysis (PCA) for product identification (Butler et al. 2016; Baranskaa et al. 2013).

10.6 X-Ray Diffraction Technique

X-ray is one of the most significant and widely used analytical technique to characterize and investigate the crystal structure by interpreting the diffraction pattern formed by scattering of x-rays by atoms, molecules and ions of crystalline structure. X-rays are considered to be the primary source for interaction with matter. X-ray diffraction is scattering of radiations (X-rays) from regular array of atoms. But X-rays can also be generated from the specimen by irradiating with electrons and alpha particles. Depending on the energy values, electrons can be emitted from the surface or form the core shells by means of several electron transitions. Surface electrons can be emitted from a specimen or inside from the core shells by means of several electron transitions like L-K shells, M-L shells etc., depending on energy available. The energy of emitted x-rays is equal to the difference between the binding energies of the electrons involved in the transition.

In X-ray diffraction, the spacing between atoms and planes of atoms is on the order of the wavelength of the x-rays. X-ray diffraction is based on Bragg's Law

$$n\lambda = 2\mathrm{dSin}\theta$$

Where n is an integer, representing the set of planes responsible for diffraction to occur, λ is the radiation wavelength, d is the spacing between the crystal planes and θ is diffracting angle that incident beam made with diffracting planes. The spacing between planes of atoms essentially functions as a three-dimensional diffraction grating. Crystalline solids show long-range periodic structure, exhibited as Bragg diffraction peaks. The Bragg peaks are symmetrically distributed diffraction patterns of focused points. Direction and intensity of diffraction depends on the crystal lattice orientations.

The Cu–Kα (0.15416 nm) source is used for most XRD analyses. Traditional x-ray diffraction is quite suitable to investigate the structural analysis of bulk crystals. Nanocrystals exhibit short coherence length of structure and a range of periodicity. This results in diffuse peaks in typical X-ray pattern. Extra special techniques need to be applied. High-energy x-ray diffraction (HEXRD) and atomic pair distribution function (PDF) analysis are found to be more appropriate for determining nanomaterials structures. Nanocrystals in powder form are suitable conducting XRD analysis (Petkov et al. 2007; Wen et al. 2005).

Small Angle X-ray Scattering Analysis. SAXS is appropriate for the particle size of the order of 1–100 nm. It is based on the scattering of x-rays by crystalline and amorphous materials have uniformly sized smaller particles. SAXS is applicable to

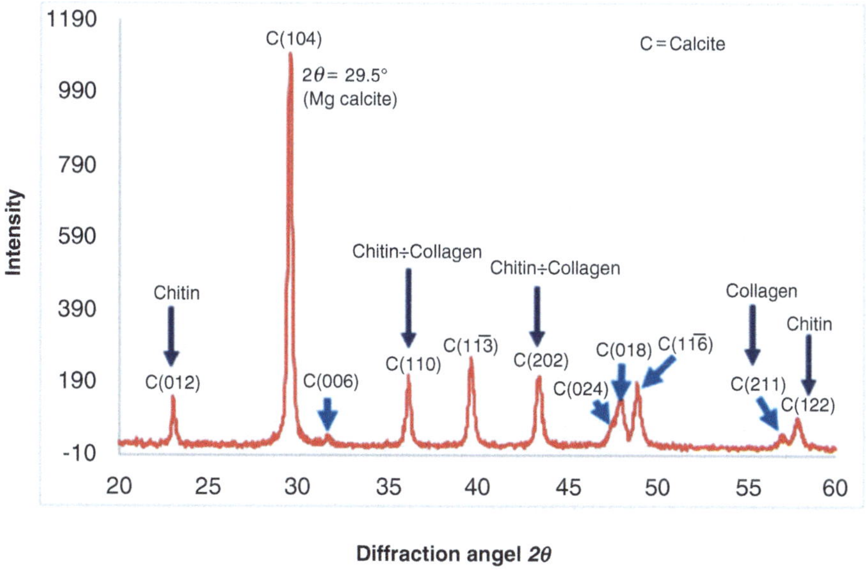

Fig. 10.11 X-ray absorption spectrum for calcified skeletal powder specimen of *C. compactum*: The 2θ scan identifies the mineral form of calcium carbonate with calcitic crystal planes, which were nucleated by chitin and collagen matrices. Black arrows show the chitin and collagen bands (Rahman 2014)

both powders specimen in the dry state and suspended specimen in a medium. Particle size can be determined by analyzing the width of Bragg peaks in x-ray diffraction spectra (Hornyak et al. 2008).

A diffraction spectrum can be analyzed by comparing the diffraction angles of corresponding peaks in the spectrum and their intensities with the diffraction standards (for example JCPDS file). Now a day this technique is being modified by using computer software to investigate and interpret the date for atomic structures and particle sizes investigation. Thus, the diffraction pattern is considered to be as "Finger print" of crystal structure to identify the phases. Spectrum peak should be symmetrical and sharp, broad peaks depict the poor quality of crystal, lattice imperfections, crystalline size and equipment effect. For example, x-ray spectrum identifies the mineral form of calcium carbonate with crystal planes nucleated by chitin and collagen matrix (Fig. 10.11).

Comparative intensities of different diffraction peaks are expressed as percentage intensity of shortest diffracted peak, is an important to identify unknown crystal phases and to refine the structure models (Seligman et al. 1966). X-ray diffraction is quite useful in bio-chemistry, solid state physics, biophysics as well as biology to analyze 3D complex structures. This technique acts as bridge between physics, chemistry and biology.

References

Amzallag A, Vaillant C, Jacob M, Unser M, Bednar J, Kahn JD, Dubochet J, Stasiak A, Maddocks JH (2006) 3D reconstruction and comparison of shapes of DNA minicircles observed by cryo-electron microscopy. Nucleic Acids Res 34(18)

Baram M, Kaplan WD (2008) Quantitative HRTEM analysis of FIB prepared specimens. J Microscopy 232(3):395

Baranskaa M, Maciej R, Jan CD, Hartwig S, Rafal B (2013) Recent advances in raman analysis of plants: alkaloids, carotenoids, and polyacetylenes. Curr Anal Chem 9:108–127

Barmparis GD, Zbigniew L, Nuria L, Remediakis IN (2015) Nanoparticle shapes by using Wulff constructions and first-principles calculations. Beilstein J Nanotechnol 6:361–368

Barnes PRF, Mulvaney R, Wolff EW, Robinson KA (2002) A technique for the examination of polar ice using the scanning electron microscope. J Microscopy 205(2):118–124

Binnig G, Rohrer H, Gerber C, Weibel W (1982) Surface studies by scanning tunneling microscope. Phy Rev Lett 49:57

Blackie EJ, Le R, Eric C, Meyer M, Etchegoin PG (2007) Surface enhanced raman scattering enhancement factors: a comprehensive study. J Phys Chem 37:13794–13803

Boyle WS, Smith GE (1970) Charge-coupled semiconductor devices. Bell Sys Tech J 49:587–593

Brandon D, Kaplan W D (2008). Microstructural characterization of materials, second edition, Chapter 5: Scanning electron microscopy, John Wiley & Sons Ltd, USA 185.

Brożek-Mucha Z (2014) Scanning electron microscopy and X-ray microanalysis for chemical and morphological characterization of the inorganic component of gunshot residue: selected problems. Biomed Res Int. Article ID 428038

Burgess J (1987) Under the microscope: a hidden world revealed. CUP Arch 11

Butler HJ, Lorna A, Benjamin B, Gianfelice C, Kelly C, Jennifer D, Karen EW, Nigel JF, Benjamin G, Pierre LMH, Michael JW, Martin RMA, Nicholas S, Francis LM (2016) Using Raman spectroscopy to characterize biological materials. Nat Prot 11:664–687

Danilatos GD (1990) Theory of the gaseous detector device in the ESEM. Adv Elect Electron Phys 78:1–102

Donovan JJ (1992) In: Goldstein JI, Newbury DE, Echlin P, Joy D C, Fiori C, Lifshin E (eds) Scanning electron microscopy and x-ray microanalysis, 2nd edn, Plenum Press, New York

Ellisman MH (2005) An historical perspective on digital imaging in transmission electron microscopy: Looking into the future. Microscopy Microanal 11:604–605

Goldstein GI, Newbury DE, Echlin P, Joy DC, Fiori C, Lifshin E (1981) Scanning electron microscopy and x-ray microanalysis. Plenum, New York

Goldstein JI, Newbury DI, Echlin P, Joy DC, Fiori C, Lifshin E (1992) Scanning electron microscopy and x-ray microanalysis, 2nd edn. Plenum Press, New York

Hindmarsh JP, Russell AB, Chen XD (2007) Fundamentals of the spray freezing of foods—microstructure of frozen droplets. J Food Eng 78(1):136–150

Hornyak GL, Tibbals HF, Joydeep D, John JM (2008) Introduction of nano science and nanotechnology, Chapter No: 3. CRC Press, Boca Raton

Hortolà P (2010) Using digital color to increase the realistic appearance of SEM micrographs of bloodstains. Micron 41(7):904–908

Hull D, Bacon J (2001) Introduction to dislocations, 4th edn. Butterworth-Heinemann, Oxford

Jeffree CE, Read ND (1991) Ambient- and Low-temperature scanning electron microscopy. In: Hall JL, Hawes CR (eds) Electron microscopy of plant cells. Academic, London, pp 313–413

Karnovsky MJ (1965) A formaldehyde-glutaraldehyde fixative of high osmolality for use in electron microscopy. J Cell Biol 27(2):137A

Keirnan JA (2000) Formaldehyde, formalin, paraformaldehyde and glutaraldehyde: What they are and what they do. Microscopy Today 1:8–12

Kelsall RW, Hamley IW, Geoghegan M (2005) Nanoscale science and technology. Wiley, London

Kirkland E (1998) Advanced computing in electron microscopy. Springer, Heidelberg, pp 19–39

Klein T, Buhr E, Frase CG (2012) TSEM: A review of Scanning Electron Microscopy in transmission mode and its applications. Adv Imaging Electron Phys 171:297–356

Liu F, Junshu W, Kunfeng C, Dongfeng X (2010) Morphology study by using scanning electron microscopy. In: Méndez-Vilas A, Díaz J (eds) Microscopy: science, technology, applications and education. Formatex Research Center, Badajoz, pp 1781–1792

Malick LE, Wilson RB, Stetson D (1957) Modified thiocarbohydrazide procedure for scanning electron microscopy: routine use for normal, pathological, or experimental tissues. Biotech Histochem 50(4):265–269

Mancuso JF, Maxwell WB, Danilatos GD (1998) Secondary electron detector for use in a gaseous atmosphere. US patent 478518, issued at 11–15, 1998

Manzer HS, Mohamed HA, Mohammad F, Mutahhar YA (2015) Chapter 2: role of nanoparticles in plants, Springer, New York

Mehta A (2012). Ultraviolet-Visible (UV-Vis) spectroscopy – derivation of Beer Lambert law in analytical chemistry. Available from: http://pharmaxchange.info/press/2012/04/ultraviolet-visible-uv-vis-spectroscopy-derivation-of-beer-lambert-law

Nawa Y, Wataru I, Sheng L, Yoshimasa K, Susumu T, Chia-Yi F, Huan-Cheng C (2014) Multicolor imaging of fluorescent nanodiamonds in living HeLa cells using direct electron-beam excitation. Chem Phy Chem 15:721–726

Petkov V, Ohta T, Hou Y, Ren Y (2007) Atomic-scale structure of nanocrystals by high-energy x-ray diffraction and atomic pair distribution function analysis: Study of FexPd100–x (x = 0, 26, 28, 48) nanoparticles. J Phys Chem C 111:714–720

Phillips R (1961) Diamond knife ultra microtomy of metals and the structure of microtomed sections. Bri J Appl Phys 12(10):554

Policarp H (2005) SEM examination of human erythrocytes in uncoated bloodstains on stone: use of conventional as environmental-like SEM in a soft biological tissue and hard inorganic material. J Microscopy 218(2):94–103

Rahman MA (2014) First evidence of chitin in calcified algea: new insights into the calcification process of *Clathromorphum compactum*. Sci Rep 4:6162

Scalf J, West P (2007) Introduction to nanoparticle characterization with AFM, Pacific Nanotechnology, Inc., www.nanoparticles.pacifi cnanotech.com

Schulz H (2014) Qualitative and quantitative FT-Raman analysis of plants, chapter 9. In: Baranska M (ed) Optical spectroscopy and computational methods in biology and medicine. Springer, Dordrecht

Seligman AM, Hannah LW, Hanker JS (1966) A new staining method for enhancing contrast of lipid-containing membranes and droplets in osmium tetroxide-fixed tissue with osmiophilic thiocarbohydrazide (TCH). J Cell Biol 30(2):424–432

Spence JCH, Zhou JM (1998) Large dynamic range, parallel detection system for electron diffraction and imaging. Rev Sci Instrum 59:2102

Sugunan A, Petter M, Johan S, Jons GH, Joydeep D (2007) Nutrition-driven assembly of colloidal nanoparticles: Growing fungi assemble gold nanoparticles as microwires. Adv Mater 19:77–78

Suzuki E (2002) High-resolution scanning electron microscopy of immunogold-labelled cells by the use of thin plasma coating of Osmium. J Microscopy 208(3):153–157

Tay MZM, Nithin SK, Gabor LH, Joydeep D (2013) Hydrophobic/ hydrophobic switching on zinc oxide micro-textured surface. Appl Surf Sci 264:344–348

Villanueva-Amadoz U, Alessandro B, Jesús M, Luis MS, José BD (2012) Focused ion beam nano-sectioning and imaging: a new method in characterisation of palaeopalynological. Grana 51:1–9

Wen X, Wang S, Ding Y, Wang ZL, Yang S (2005) Controlled growth of largearea, uniform, vertically aligned arrays of -fe2o3 nanobelts and nanowires. J Phy Chem B 109:215–220

Wergin WP, Erbe EF(1994) Snow crystals: capturing snow flakes for observation with the low-temperature scanning electron microscope. Scanning 16, 17(1):41–50

Wittke JH (2006) Effects of electron bombardment. Northern Arizona University. https://nau.edu/microanalysis/Microprobe/Interact-Effects.html

Zach P, Sharon H, Sofiya K, Porat Z, Yehuda Z (2014) Green synthesis of gold nanoparticles using plant extracts as reducing agents. Int J Nanomed 9:4008–4021

Chapter 11
Future Prospects of Nanobotany

Sumera Javad, Madeeha Ansari, and Iqra Akhtar

11.1 Introduction

Synthesis of nanoparticles by biological methods has been developed for last few decades. Biological methods never required toxic chemicals therefore these have grown fast (Konwarh et al. 2011). Biological methods employ micro-organisms such as cell culture of bacteria, fungi as well as plants. All these used methods are simple, eco-friendly and cost effective for synthesis of nanoparticles (Rauwel et al. 2015). But plant extracts are considered as the good and easy medium for synthesis of nanoparticles. The usage of plant extracts has eliminated the difficult downstream processes and controlling of cell culture (Shankar et al. 2004). The nanoparticles formation is faster in plants as compared to microorganisms due to better reduction of metal to nanoparticles (Kumar et al. 2012). The use of plant extracts are also preferred than to the microorganisms due to bio safety reasons (Kumar and Yadav 2009).

11.2 Nanobotany in Agriculture

The research for application of nanotechnology has been ongoing for many years to solve agricultural issues as well as problems of environment such as agrochemicals sustainability, climatic change, improvement of crop varieties and increase in production. Researchers have proved potential of nanotechnology in the field of agriculture especially to protect the crops form pests and in disease management (Sastry et al. 2010; Gogos et al. 2012). The nanomaterials usage in agriculture can reduce the amount of sprayed agrochemicals and minimize nutrient loss by smart delivery of active ingredients (Gogos et al. 2012).

S. Javad (✉) · M. Ansari · I. Akhtar
Department of Botany, Lahore College for Women University, Lahore, Pakistan

S. Javad, A. Butt (eds.), *Nanobotany*, https://doi.org/10.1007/978-3-319-77119-9_11

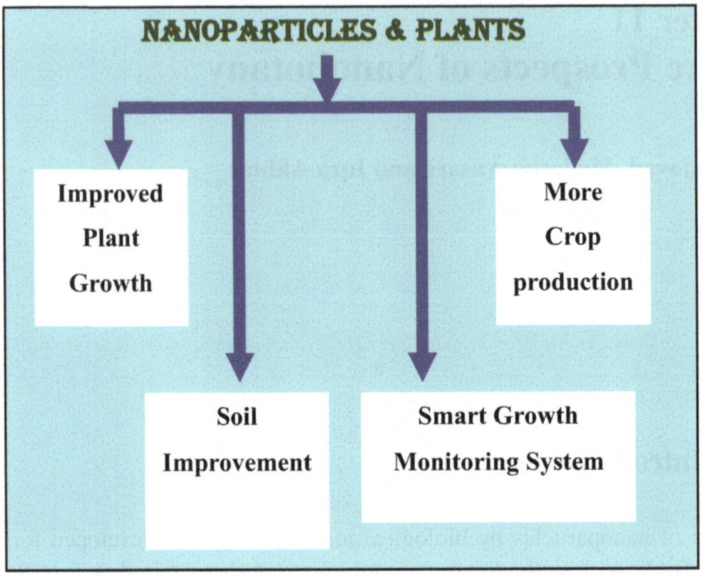

Fig. 11.1 Uses of nanoparticles in agriculture

The yield of crop can be increased by optimizing the water and nutrient content. The devices which are based on nanotechnology are being used in the field of genetic transformation and plants breeding (Torney et al. 2007). Nanoparticles have potential to enhance the efficiency of pesticides and insecticides which are much needed to control the agricultural pests. The use of nanoparticles can help to reduce the adverse effects of commonly used insecticides to the different populations such as human beings, plants, animals and micro-organisms of a community. It can be used at commercial level in order to maximize the safety level of environment and protection of human health as well. Formulation of commercial products will be combination of the botanical insecticides with nanoparticles in near future (De Oliveira et al. 2014).

There are multiple research reports about the growth promoting role of nanoparticles in plants. They increase biomass production of plants like iron oxide, zinc oxide and copper oxide nanoparticles (Burman et al. 2013; Dhoke et al. 2013; Dimpka et al. 2012). They have also been reported to increase shoot length, root length (Juhel et al. 2011), flower size (Khodakovskaya et al. 2013), chlorophyll content (Mahmoodzadeh et al. 2013) and seed vigor (Morla et al. 2011) (Fig. 11.1).

11.2.1 In Controlling Agrochemicals (Pesticides)

Sustainable production of food for increased human population is the serious challenge for agricultural sector (Godfray and Garnett 2014; Mcclung 2014). It has become necessary to maximize the productivity. There is an increased use of fertilizers and pesticides. Increased use of such pesticides causes severe damage to humans and other life forms. The harmful effects of these pesticides are dependent upon toxicity level, the degree of contamination and the duration of exposure (Köhler and Triebskorn 2013). The amount of pesticides which are practiced in the fields over crops reached to 2.5 million tons each year (Fenner et al. 2013). This over practice of pesticides has affected the food quality, caused the pests to become resistant, generated considerable waste causing pollution in environment and danger to public health (Abhilash and Singh 2009; Köhler and Triebskorn 2013). It has been reported that pesticides which are being used causes pollution in environment. These are lost more than 90% to the air and water as a result of runoff (Stephenson 2003; Ghormade et al. 2011). All these issues have promoted the need of protected and efficient pesticide application methods to prevent the environment from adverse effects of pesticides. Nanotechnology provided innovative tools that can be used to deliver agrochemicals safely without causing disturbance in ecosystem. It provided such carriers systems that can control the release of agents for pest at specified place. It can reduce the amount of pesticides being concentrated in the environment (Ghormade et al. 2011; González et al. 2014). Nanosensors can be used to monitor the conditions of environment. Interactions between plants and microbes can also be detected with these sensors to prevent from the predicted damage (Nair et al. 2010, Ghormade et al. 2011). There are several nanometric systems such as nanoparticles, nanocapsules and viral capsids can be used as a smart delivery system for active compounds to control pests and diseases (Anjali et al. 2012; Kah et al. 2013; Campos et al. 2014; Kah and Hofmann 2014). NPs have larger surface area and higher rate of mobility due to enough smaller size. So, these can improve the efficiency of agrochemicals. The level of toxicity can also be decreased due to reduction in use of organic solvents and commercially formulated conventional pesticides (Sasson et al. 2007).

11.2.2 Nanofertilizers

Fertilizers play an important role for growth and development of plant. But most of the times applied fertilizers are not available to plants because of many issues such as decomposition, degradation, leaching and water logging. Due to which plants become nutrients deficient and cannot give better yield. Therefore, it is necessary to decrease the nutrients loss. The yield of crop can be increased by using tools of nanotechnology and nanomaterials. Nutrients can be applied in the form of nano fertilizers and nanocapsules that means minerals of nanosizes. These will be more effective to give better yield. These nano tools will release the nutrients on-demand. These can also control the release of chemicals and nutrients on specific target sites important for growth of plants (Derosa et al. 2010; Nair et al. 2010). In literature it is illustrated that nanoparticles can improve metabolism of plants due to their physiochemical properties (Giraldo et al. 2014). Nanoparticles show positive effects in stimulating plant growth. Since, micro element deficiency is very much common in cereals and millions of hectares of the world's arable land are facing a shortage of one or more micro nutrients (Lovely et al. 1991). The concept of nanoparticles is in process of development and can increase the interest of the industry in the phenomenon with the passage of time. The need of the hour is to manage the research in typical related topics like which nanoparticles can be used as nanofertilizers safely? How can they be produced in bulk amounts? What is their lifespan? It also needs to study the specific nanofertilizers with specific crop demands and study of any side effects if they are (Christian and Prem 2017).

11.3 Nanobotany in Medicine

Medicine is the field growing on with the increasing population and diseases. Every day there are research reports being submitted about discovery of some new medicine or may be some new parasite and even some diverse form of a pre-existing disease or syndrome. In some parts of the world, there is also a high threat of antibiotic resistant strains of microbes which is a real challenge to medical scientists. Nanotechnology has come across with some new aspects and it seems to have potential to combat these threats.

11.3.1 In Diagnosis of Diseases

Cure of the diseases is dependent on the visible symptoms which appear at some stage of the disease. Before appearance of symptoms professionals can never recognize the specific illness with which patient suffers. It is necessary to detect the disease at earlier stage for a better chance to cure it particularly in case of fatal diseases like cancer. Diagnosis of nucleic acid can play an important role as it allows the detection of disease at early stage where cells are symptomless mostly. Polymerase chain reaction (PCR) technology offers such assays and devices which help in detection. But nanotechnology is offering such tools as semiconductor nanocrystals (quantum dots) which can help efficiently with greater sensitivity and cost effective economically. These use probes which can resist to damage significantly during excitation and light emission cycles more than organic molecules and help in detection of fault (Drexler 1992).

11.3.2 In Delivery of Drugs

Drugs can be delivered to the specific sites with the help of nanoparticles that is not quite easy for standard drugs. Nanoparticles can guide the chemically attached drugs to reach specific diseased or infectious site by radio or magnetic signals. Drugs encapsulated in packing's made with the nano materials such as nanoshells, dendrimers and hollow polymer capsules can be used to avoid the side effects of potent drugs. The amount of the delivered drug can also be controlled. Release of such drugs can be controlled more precisely in the presence of specific molecule or external triggers (like infrared heat) are given. These nanodrugs can be controlled and released at a time when specific molecule or external triggers (like infrared heat) are given (Lavan et al. 2002). Nanotechnology can help in administering such drugs which can be taken orally due to poor bioavailability (El-Shabouri 2002; Hu et al. 2004). Nanotechnology can also help in delivery of antigens for vaccination (Diwan et al. 2003; Köping-Höggård et al. 2005). Microparticles and nanoparticles can also enhance the immunization (Lutsiak et al. 2002).

11.3.3 In Gene Delivery

The implementation of gene therapy is suffering with various difficulties. There is always a possibility that mutant type will be generated during gene therapy and can't be reversed to its original form. Viral vectors cause problems while transferring the genes (Young et al. 2006). But nonviral vectors based on nanoparticles (size ranges from 50–500 nm) can be useful to transport the plasmids. Nanosized gene carriers are less immunogenic and helpful during repairing or replacement of human genes (Davis 1997). Liposomes can pass through the lipid bilayer due to its similar composition. Therefore, these are used in gene therapy. But it has been noticed that some groups of liposomes causing problem in local delivery of genes (Ewert et al. 2005; Hart 2005). It is reported that liposomes treated with nanoparticles (polyethylene glycol) used to express the antibody for human insulin in the brains of rhesus monkeys. Success of experiment marked the importance of nanotechnology for the future of targeted gene therapy and in advancement of molecular medicine (Zhang et al. 2003).

11.3.4 In Biopharmaceuticals

Nanobotany can help in development of such drugs that pharmaceuticals cannot achieve easily. The pharmaceutical industries mainly focus on development of drugs which are used to treat about 500 confirmed diseases. Discovery of drugs can be cost effective. It can profile large number of compounds. Nanoscale tools will be a great favor for the development of new drugs especially to small medicine industries which cannot afford wastage of thousands of compounds and pay to hundreds of organic chemists to synthesize drugs. Nanotechnology has ability to target the compounds on solid substrates. It can control the time and place of reactions. It can control the amount of required reagents and solutions in a process. It can help in production of specified drugs. Potentia Pharmaceuticals (an early-stage company) is using the nanotechnologies to develop the new drugs (Lavan et al. 2002).

11.3.5 In Cardiac Therapy

Nanotechnology is offering such tools which can help to explore problems at the cellular level. These tools can be used to treat challenges of cardiovascular diseases effectively. These can be helpful to diagnose and imaging in the field of tissue engineering (Wickline et al. 2006). Tools such as nanosensores, nanocrystals, quantum dots (QDs) and nanobarcodes can be used to sense and monitor the response of complex immune system to cardiac or inflammatory issues (Guccione et al. 2004). Mechanisms which are significant and clinically implicated to treat cardiac disorders can be detected and described with the help of this new emerging technology. It can be used to design nanomachines and biological systems can be treated at the molecular level with these. Such devices can change the treatment ideas and methods which are being used to treat cardiovascular diseases. It will be a great achievement in the field of cardiac therapy (Panyam and Labhasetwar 2003).

11.3.6 Dental Care

Nanobotany will have bright future in the field of dentistry by using nanomaterials (Shi et al. 1999; West and Halas 2000). Nanorobots can provide rapid and reliable repair of damaged teeth or to maintain the teeth health. These can be used to treat the specific tubules of teeth within minutes. It can also help in repairing the tooth disorders and fill the cavities with nano materials like nanoparticles rods of calcium hydroxypatatite crystals (Shellhart and Oesterle 1999).

11.4 Nanobotany in Construction Industry

Products generated by nanotechnology can be used to design and construct the infrastructures in construction industry. These products have enormous properties such as lighter in weight, strong structural composition (can be used to build up the bridges), improve joining property of materials, reduce the heat transfer rate of fire insulators and increase the reflectivity of glass (Rana et al. 2009).

11.4.1 In Concrete

Concrete is the construction material used most commonly all over the world. Now, by nano-scale the size and the properties of cementation materials can be studied. It has been reported after thorough research work that reactions of alkali and silicate, process of hydration and reactivity of fly ash are affected by the use of nanotechnology (Balaguru 2005). Adhesive and strength properties of cements can be improved by adding nanoscale materials. Scientist found that silicon nanoparticles help in increasing the compressive strength of concrete, having large amount of fly ash and fill the pores present between larger size fly ash and particles of cements (Li 2004). Cracking in the constructed areas is one of the major issues. Scientists are working on producing healing polymers which are based on nanotechnology. These polymers include nano-encapsulated healing agent along with a catalyst. These microcapsules are broken when incorporated into cracks due to which healing agent released and bind with the used catalyst. The polymerization occurs immediately over there and fills the cracks. The cracking present in the bridge areas can be fixed by self healing polymer. But it requires costly epoxy injection (Bigley and Greenwood 2003).

11.4.2 In Structural Composites

Steel is one of the materials widely used in construction industry for strength. It has great properties, very important for designing and construction, such as strength, corrosion resistance and weld ability. Scientists have developed the low carbon high performance steel in 1992 to construct the long bridges. The new developed steel has more ability to resist against the corrosion and weld easily when copper NPs are incorporated. Sandvik Nanoflex Materials Technology is a company making new type of stainless steel that has higher strength and a good surface. This is suitable to use especially in making the lightweight and stiff structures. It has "plywood" resembling different structures in comparison with the commonly used conventional steel. These microstructures make the steel stronger, stiff, ductility, and more corrosion resistant (Arafa et al. 2006).

11.4.3 In Coating

Nanoparticles/nanolayers incorporated coatings have been developed for certain purpose. TiO_2 can be utilized to coat glazing due to its properties such as sterilizing and anti fouling. TiO_2 can break and disintegrate the organic dirt by powerful catalytic reaction. It is hydrophilic in nature and allows the spreading of water over the surface evenly that's why help to wash away dirt. Other special coatings also have been developed, such as thermal control, energy sawing and anti reflection coating (Arafa et al. 2006). It has a vast horizon to develop new nanocoating materials with advanced characters to cope with new environmental issues.

11.5 Nanobotany in Food Industry

Nanotechnology has great participation in food industry as well. It has history from onward when Pasteur introduced the Pasteurization process to kill the bacteria. It was the first revolution in improvement of food quality. Nanotechnology is now invading the food industry and establishing great potential. Food industry has a number of applications of Nanotechnology which include increasing the flavor, encapsulation of food particles to control their digestibility and delivery of substances in targeted sites, antimicrobial effect enhancement in food products thus increasing their shelf life and making food storage better. The color and taste of food items can be changed by using nanoparticles and can also remove the bacteria. Nanobarcodes are also in use for safety and to monitor distribution of food products (Chellaram et al. 2014).

11.5.1 In Food Processing

In food industry, food processing is an important step in which raw ingredients are converted to form the food products. This processed food has longer shelf life than to the normal. This process includes the steps such as removal of toxic chemicals, prevent from pathogens, preserve for longer time and improve the consistency for better distribution. Foods which are processed usually become less susceptible to pathogens than fresh food material. Therefore, these can be transported to longer distance easily. Incorporation of the nanotechnology in food sciences is improving the processing of foods now a day. Nanocapsules contribute to food processing. Functional properties of food are maintained by encapsulating it into simple solutions, colloids, emulsions and biopolymers (Abbas et al. 2009). Nanodrops are nano sized lipids which served as liquid carrier of healthy components that are insoluble in water and fats. These are helpful to inhibit transportation of cholesterol from the digestive system into the bloodstream (Dingman and Rehs 2008).

11.5.2 In Nutritional Supplements

Nanofood's total market value is reached to US$5.8 billion, out of which food ingredients consumes US$1475 million, food processing US$1303 million, food safety US$97 million and food packaging US$2.93 billion, according to 2012 survey (Sekhon 2010). Nanoceuticals are commonly used commercial supplements. Nanosized materials are also used to increase the nutrients absorption. Nanocochleates are very much effective tools for delivery of nutrients to cells and there is no effect on color, aroma and flavor of food products. Food supplements produced with the help of nanotechnology are more effective than commonly used supplements because due to smaller size they react more efficiently (Avella et al. 2005; Chaudhry et al. 2008; Alfadul and Elneshwy 2010).

11.6 Future of Nanobotany in Other Fields

11.6.1 In Nano Cloak Formation

It has been shown that an object can be concealed into a cloak composed of suitable metamaterial (MMs) from observer (Schurig et al. 2007). Efficiency of a cloak depends on some controlled properties of material such as permittivity and permeability. Implementation of these controlled parameters is not easily accessible in natural media. MMs having optical properties are artificial, so these can be controlled by controlling unit cells (Schurig et al. 2006). Then formation of cloak by using elliptical metallic nanoparticles (NPs) was also implemented. The elliptical shape of metallic nanoparticles varies due to change in dimensions across the radius. Wherever the required radial and azimuthal permittivity acquired across the cylindrical cloak, it become possible to hide the objects from surrounding. This can only be achieved if the material forming the cloak may be considered as an effectively homogenous medium with tunable properties (Cai et al. 2007). This homogenous mixture of NPs can be made easily formed from the plant extracts. The formation of NPs from plant extracts is quite easy as mentioned in first paragraph of the chapter.

11.6.2 In Nanorobotics

Nanorobotics is another application of nanotechnology getting higher interest with the passage of time. It seems like nanorobots will not be only in the science fiction for much longer. Nanorobots have multiple dimensions like the biological cells. These nanorobots can help in monitoring of environment for microorganisms. These can also aid in caring human health such as enhance the circulatory system, detect small concentrations of pathogens and destroy them. Nanorobots can repair the cells due to its small size enough to penetrate in the cell (Vettiger et al. 2002). Nanorobots are useful in medical filed as well as other fields such as help in formation of controlled surfaces of variable friction. Construction of artificial nanorobots is quite difficult due to difficulties in building nanostructures (Requicha 2003). It will not be more difficult for scientists to construct the nanostructures for much longer. Nanoparticles of controlled properties can be formed from plant extracts. These can be produced in large amount, will be cost effective as plants are the natural resources and abundantly present in our surroundings. Plants are renewable source for the synthesis of metal nanoparticles. Formation of nanostructures for nanorobots will be easy to construct with these controlled NPs.

References

Abbas KA, Saleh AM, Mohamed A, Mohdazhan N (2009) The recent advances in the nanotechnology and its applications in food processing: a review. J Food Ag Env 7(3–4):14–17

Abhilash P, Singh N (2009) Pesticide use and application: an Indian scenario. J Haz Mat 165(1):1–12

Alfadul S, Elneshwy A (2010) Use of nanotechnology in food processing, packaging and safety–review. Afr J Food Agr Nut Develop 10(6)

Anjali C, Sharma Y, Mukherjee A, Chandrasekaran N (2012) Neem oil (*Azadirachta indica*) nanoemulsion—a potent larvicidal agent against *Culex quinquefasciatus*. Pest Manag Sci 68(2):158–163

Arafa MD, Defazio C, De Miguel Y, Porro A, Bartos P (2006) Nano-composite coatings for transportation infrastructures: demonstration projects. In: NICOM 2: 2nd international symposium on nanotechnology in construction, 2006. RILEM Publications SARL, pp 363–371

Avella M, De Vlieger JJ, Errico ME, Fischer S, Vacca P, Volpe MG (2005) Biodegradable starch/clay nanocomposite films for food packaging applications. Food Chem 93(3):467–474

Balaguru P (2005) Nanotechnology and concrete: background, opportunities and challenges. In: Applications of nanotechnology in concrete design: proceedings of the international conference held at the University of Dundee, Scotland, UK. Thomas Telford Publishing, pp 113–122

Bigley C, Greenwood P (2003) Using silica to control bleed and segregation in self-compacting concrete. Concrete 37(2):43–45

Burman U, Saini M, Kumar P (2013) Effect of zinc oxide nanoparticles on growth and antioxidant system of chickpea seedlings. Toxicol Environ Chem 95(4):605–612

Cai W, Chettiar UK, Kildishev AV, Shalaev VM (2007) Optical cloaking with metamaterials. Nat Photonics 1(4):224–227

Campos EVR, De Oliveira JL, Fraceto LF (2014) Applications of controlled release systems for fungicides, herbicides, acaricides, nutrients, and plant growth hormones: a review. Adv Sci Eng Med 6(4):373–387

Chaudhry Q, Scotter M, Blackburn J, Ross B, Boxall A, Castle L, Aitken R, Watkins R (2008) Applications and implications of nanotechnologies for the food sector. Food Addit Contamin 25(3):241–258

Chellaram C, Murugaboopathi G, John A, Sivakumar R, Ganesan S, Krithika S, Priya G (2014) Significance of nanotechnology in food industry. APCBEE Procedia 8:109–113

Christian OP, Prem SB (2017) Nanofertilizers: new products for the industry. J Agric Food Chem. https://doi.org/10.1021/acs.jafc.7b02150

Davis S (1997) Biomedical applications of nanotechnology—implications for drug targeting and gene therapy. Trends Biotech 15(6):217–224

De Oliveira JL, Campos EVR, Bakshi M, Abhilash P, Fraceto LF (2014) Application of nanotechnology for the encapsulation of botanical insecticides for sustainable agriculture: prospects and promises. Biotech Adv 32(8):1550–1561

Derosa MC, Monreal C, Schnitzer M, Walsh R, Sultan Y (2010) Nanotechnology in fertilizers. Nat Nanotech 5(2):91–91

Dhoke SK, Mahajan P, Kamble R, Khanna A (2013) Effect of nanoparticles suspension on the growth of mung (*Vigna radiata*) seedlings by foliar spray method. Nanotechnol Dev 3(1):1

Dimkpa CO, McLean JE, Latta DE, Manangón E, Britt DW, Johnson WP, Boyanov MI, Anderson AJ (2012) CuO and ZnO nanoparticles: phytotoxicity, metal speciation, and induction of oxidative stress in sand-grown wheat. J Nano Res 14(9):1–15

Dingman J, Rehs D (2008) Nanotechnology: its impact on food safety. J Env Health 70(6):47–50

Diwan M, Elamanchili P, Lane H, Gainer A, Samuel J (2003) Biodegradable nanoparticle mediated antigen delivery to human cord blood derived dendritic cells for induction of primary T cell responses. J Drug Targ 11(8–10):495–507

Drexler KE (1992) Nanosystems: molecular machinery, manufacturing, and computation, Wiley, New York

El-Shabouri M (2002) Positively charged nanoparticles for improving the oral bioavailability of cyclosporin-A. Int J Pharm 249(1):101–108

Ewert K, Evans HM, Ahmad A, Slack NL, Lin AJ, Martin-Herranz A, Safinya CR (2005) Lipoplex structures and their distinct cellular pathways. Adv Genetics 53:119–155

Fenner K, Canonica S, Wackett LP, Elsner M (2013) Evaluating pesticide degradation in the environment: blind spots and emerging opportunities. Science 341(6147):752–758

Ghormade V, Deshpande MV, Paknikar KM (2011) Perspectives for nano-biotechnology enabled protection and nutrition of plants. Biotech Adv 29(6):792–803

Giraldo JP, Landry MP, Faltermeier SM, Mcnicholas TP, Iverson NM, Boghossian AA, Reuel NF, Hilmer AJ, Sen F, Brew JA (2014) Plant nanobionics approach to augment photosynthesis and biochemical sensing. Nat Mat 13(4):400–408

Godfray HCJ, Garnett T (2014) Food security and sustainable intensification. Phil Trans R Soc B 369(1639):20120273

Gogos A, Knauer K, Bucheli TD (2012) Nanomaterials in plant protection and fertilization: current state, foreseen applications, and research priorities. J Agr Food Chem 60(39):9781–9792

González JOW, Gutiérrez MM, Ferrero AA, Band BF (2014) Essential oils nanoformulations for stored-product pest control–characterization and biological properties. Chemosphere 100:130–138

Guccione S, Li KC, Bednarski MD (2004) Vascular-targeted nanoparticles for molecular imaging and therapy. Meth Enz 386:219–236

Hart SL (2005) Lipid carriers for gene therapy. Curr Drug Del 2(4):423–428

Hu L, Tang X, Cui F (2004) Solid lipid nanoparticles (SLNs) to improve oral bioavailability of poorly soluble drugs. J Pharm Pharmacol 56(12):1527–1535

Juhel G, Batisse E, Hugues Q, Daly D, Pelt FN, O'Halloran J, Jansen MA (2011) Alumina nanoparticles enhance growth of *Lemna minor*. Aqua Toxicol 105(3):328–336

Kah M, Hofmann T (2014) Nanopesticide research: current trends and future priorities. Env Int 63:224–235

Kah M, Beulke S, Tiede K, Hofmann T (2013) Nanopesticides: state of knowledge, environmental fate, and exposure modeling. Crit Rev Env Sci Tech 43(16):1823–1867

Khodakovskaya MV, Kim BS, Kim JN, Alimohammadi M, Dervishi E, Mustafa T, Cernigla CE (2013) Carbon nanotubes as plant growth regulators: effects on tomato growth, reproductive system, and soil microbial community. Small 9(1):115–123

Köhler HR, Triebskorn R (2013) Wildlife ecotoxicology of pesticides: can we track effects to the population level and beyond? Science 341(6147):759–765

Konwarh R, Gogoi B, Philip R, Laskar M, Karak N (2011) Biomimetic preparation of polymer-supported free radical scavenging, cytocompatible and antimicrobial "green" silver nanoparticles using aqueous extract of *Citrus sinensis* peel. Coll Surf B Biointerf 84(2):338–345

Köping Höggård M, Sánchez A, Alonso MJ (2005) Nanoparticles as carriers for nasal vaccine delivery. Exp Rev Vacc 4(2):185–196

Kumar V, Yadav SK (2009) Plant-mediated synthesis of silver and gold nanoparticles and their applications. J Chem Tech Biotech 84(2):151–157

Kumar R, Roopan SM, Prabhakarn A, Khanna VG, Chakroborty S (2012) Agricultural waste Annona Squamosa peel extract: biosynthesis of silver nanoparticles. Spectrochimica Acta Part A Mol Biomol Spec 90:173–176

Lavan DA, Lynn DM, Langer R (2002) Moving smaller in drug discovery and delivery. Nat Rev Drug Discov 1(1):77–84

Li G (2004) Properties of high-volume fly ash concrete incorporating nano-SiO_2. Cement Concrete Res 34(6):1043–1049

Lovely D, Phillips EJ, Lonergan DJ (1991) Enzymatic versus nonenzymatic mechanisms for Fe (III) reduction in aquatic sediments. Env Sci Tech 25(6)

Lutsiak MC, Robinson DR, Coester C, Kwon GS, Samuel J (2002) Analysis of poly (D, L-lactic-co-glycolic acid) nanosphere uptake by human dendritic cells and macrophages in vitro. Pharm Res 19(10):1480–1487

Mahmoodzadeh H, Nabavi M, Kashefi H (2013) Effect of nanoscale titanium dioxide particles on the germination and growth of canola (*Brassica napus*). J Ornamental Hortic Plants 3:25–32
Mcclung CR (2014) Making hunger yield. Sci 344(6185):699–700
Morla S, Ramachandra RCSV, Chakrapani R (2011) Factors affecting seed germination and seedling growth of tomato plants cultured *in vitro* conditions. J Chem Bio Phys Sci B 1:328–334
Nair R, Varghese SH, Nair BG, Maekawa T, Yoshida Y, Kumar DS (2010) Nanoparticulate material delivery to plants. Plant Sci 179(3):154–163
Panyam J, Labhasetwar V (2003) Biodegradable nanoparticles for drug and gene delivery to cells and tissue. Adv Drug Del Rev 55(3):329–347
Rana AK, Rana SB, Kumari A, Kiran V (2009) Significance of nanotechnology in construction engineering. Int J Recent Trends Eng 1(4):46–48
Rauwel P, Küünal S, Ferdov S, Rauwel E (2015) A review on the green synthesis of silver nanoparticles and their morphologies studied via TEM. Adv Mat Sci Eng 2015
Requicha AA (2003) Nanorobots, NEMS, and nanoassembly. Proceedings IEEE 91(11):1922–1933
Sasson Y, Levy-Ruso G, Toledano O, Ishaaya I (2007) Nanosuspensions: emerging novel agrochemical formulations. Insecticides design using advanced technologies. Springer, Berlin
Sastry K, Rashmi H, Rao N (2010) Nanotechnology patents as R&D indicators for disease management strategies in agriculture. J Intellectual Property Rights 15:197–205
Schurig D, Mock J, Justice B, Cummer SA, Pendry JB, Starr A, Smith D (2006) Metamaterial electromagnetic cloak at microwave frequencies. Science 314(5801):977–980
Schurig D, Pendry J, Smith D (2007) Transformation-designed optical elements. Opt Express 15(22):14772–14782
Sekhon BS (2010) Food nanotechnology–an overview. Nanotech Sci Appl 3(1):1–15
Shankar SS, Rai A, Ahmad A, Sastry M (2004) Rapid synthesis of au, ag, and bimetallic au core–ag shell nanoparticles using neem (*Azadirachta indica*) leaf broth. J Coll Interf Sci 275(2):496–502
Shellhart WC, Oesterle LJ (1999) Uprighting molars without extrusion. The J Am Dental Assoc 130(3):381–385
Shi H, Tsai WB, Garrison MD, Ferrari S, Ratner BD (1999) Template-imprinted nanostructured surfaces for protein recognition. Nature 398(6728):593–597
Stephenson GR (2003) Pesticide use and world food production: risks and benefits. ACS Publications, Washington, DC
Torney F, Trewyn BG, Lin VSY, Wang K (2007) Mesoporous silica nanoparticles deliver DNA and chemicals into plants. Nat Nanotech 2(5):295–300
Vettiger P, Cross G, Despont M, Drechsler U, Durig U, Gotsmann B, Haberle W, Lantz M, Rothuizen H, Stutz R (2002) The "millipede"-nanotechnology entering data storage. IEEE Trans Nanotech 99(1):39–55
West JL, Halas NJ (2000) Applications of nanotechnology to biotechnology: commentary. Curr Opin Biotech 11(2):215–217
Wickline SA, Neubauer AM, Winter P, Caruthers S, Lanza G (2006) Applications of nanotechnology to atherosclerosis, thrombosis, and vascular biology. Arterioscler Thromb Vasc Biol 26(3):435–441
Young LS, Searle PF, Onion D, Mautner V (2006) Viral gene therapy strategies: from basic science to clinical application. The J Pathol 208(2):299–318
Zhang Y, Schlachetzki F, Li JY, Boado RJ, Pardridge WM (2003) Organ-specific gene expression in the rhesus monkey eye following intravenous non-viral gene transfer. Mol Vis 9:465–472

Index

S. Javad, A. Butt (eds.), *Nanobotany*, https://doi.org/10.1007/978-3-319-77119-9

The manufacturer's authorised representative in the EU is Springer Nature Customer Service Centre GmbH, Europaplatz 3, 69115 Heidelberg, Germany. If you have any concerns regarding our products, please contact ProductSafety@springernature.com

Printed and bound by CPI Group (UK) Ltd, Croydon, CR0 4YY
05/10/2026
02228141-0004